THINKING ABOUT
EQUATIONS

THINKING ABOUT EQUATIONS

A Practical Guide for Developing
Mathematical Intuition in the
Physical Sciences and Engineering

MATT A. BERNSTEIN, PhD
WILLIAM A. FRIEDMAN, PhD

A JOHN WILEY & SONS, INC., PUBLICATION

Published by John Wiley & Sons, Inc., Hoboken, New Jersey.
Published simultaneously in Canada.

For general information on our other products and services or for technical support, please contact our Customer Care Department within the United States at (800) 762-2974, outside the United States at (317) 572-3993 or fax (317) 572-4002.

Wiley also publishes is books in a variety of electronic formats. Some content that appears in print may not be available in electronic formats. For more information about Wiley products, visit our web site at www.wiley.com.

Library of Congress Cataloging-in-Publication Data:

Bernstein, Matt A., 1958–
  Thinking about equations : a practical guide for developing mathematical intuition in the physical sciences and engineering / Matt A. Bernstein, William A. Friedman.
       p. cm.
    Includes bibliographical references and index.
    ISBN 978-0-470-18620-6 (pbk.)
   1. Mathematical analysis. 2. Physical sciences–Mathematics. 3. Engineering–Mathematics. I. Friedman, William A. (William Albert) II. Title.
    QA300.B496 2009
    515—dc22

                                                                                    2009001787

10 9 8 7 6 5 4 3 2 1

CONTENTS

PREFACE

Equations play a central role in day-to-day problem solving for the physical sciences, engineering, and related fields. Those of us who pursue these technical careers must learn to understand and to solve equations. It is often a challenge to figure out what an equation means and then to form a strategy to solve it. One aim of this book is to illustrate that by investing a little effort into thinking about equations, your task can become easier and your solutions can become more reliable.

On a deeper level, studying the topics presented here could provide you with greater insight into your own field of study, and you may even derive some intellectual pleasure in the process. Today, a staggering amount of knowledge is readily available through the Internet. With this abundance of information comes an ever-greater premium on understanding the interconnections and themes embedded within this information. This book can help you gain that intuitive understanding while you acquire some practical mathematical tools. We hope that our selection of topics and approaches will open your eyes to this useful and rewarding process.

One might ask why we wrote a book about equations, i.e., focusing on analytic methods, when today so many problems are solved using numerical methods. In our own scientific research, we rely heavily on numerical computation. Equations, however, are the basis for almost all numerical solutions and simulations. A deeper understanding of the underlying equations often can improve the strategy for a numerical approach, and importantly, it helps one to check whether the resulting numerical answers are reasonable.

With the advent and widespread availability of powerful computational packages such as MATLAB®, MAPLE™, IDL®, and others, some students have spent less time studying and thinking about equations and have not yet developed an intuitive "feel" for this topic that experience brings. This book can serve as a resource and workbook for students who wish to improve those skills in an efficient manner. We believe that the powerful graphics capabilities of those computational packages can complement the material presented here, and users of symbolic manipulation software such as Mathematica® will also find this book relevant to their work.

The core of our intended audience is undergraduate students majoring in any discipline within engineering or any of the physical sciences including physics, geology, astronomy, and chemistry, as well as related fields such as applied mathematics and biophysics. Even students in peripherally related fields, such as computational finance, will find the book useful, particularly when they apply methods that originated in the physical sciences and engineering to their own work. Graduate students with an undergraduate background in a field like biology who are transitioning into a more quantitative area such as biomedical engineering could also find the techniques described in this book quite useful.

We have tried to target the level of the book so it is accessible to undergraduate students who have a background in basic calculus and some familiarity with differential equations. An undergraduate course in introductory physics is also recommended. Some of the worked example problems draw on concepts and results from linear algebra and complex analysis, but this background is not essential to follow most of the material. We are not mathematicians, and there are no formal mathematical proofs in the book. Instead, the derivations, where provided, are presented at a level of rigor that is typical of work in the physical sciences. A few of the sections and example problems are labeled with an asterisk (*), because either they assume a higher level of mathematical background or else the material contained in them is not central to the rest of the book. These starred sections and problems can be skipped without loss of continuity.

This book is *not* a comprehensive "how-to" manual of mathematical methods. Instead, selected concepts are introduced by a short discussion in each section of each chapter. Then, the concepts are illustrated and further developed with example problems followed by detailed solutions. These worked example problems form the backbone of this book.

Our own backgrounds are in physics, with specialization in magnetic resonance imaging, medical physics, and biomedical engineering (MAB) and theoretical nuclear physics (WAF). While we have chosen example problems drawn from subjects that are familiar to us, we have tried to avoid problems that require highly specialized knowledge. We also have tried to provide explanations of terms that might be unfamiliar and to avoid the use of jargon. We hope you will read and work through example problems that are *outside* of your immediate area of interest. If you can recognize concepts that can be applied to your own work, then you will have made great progress toward our ultimate goal. At the end of each chapter, there are a number of exercises designed to test and further develop your understanding of the material. We also encourage you to devise and to solve some of your own exercises, perhaps related to your own specific field of interest.

Chapter 1 deals with units and dimensions of physical quantities. Although this is a very basic topic, it has some surprisingly profound implications that are developed more fully in Chapter 6, which deals with dimensional analysis and scaling. Chapter 2 illustrates several common pitfalls to avoid when dealing with equations, along with a few handy techniques and tricks that are used in later chapters. Chapter 3 discusses special cases and their use to check equations and to guess at the solution of difficult problems. Chapter 4 focuses on pictorial and graphical methods, and provides a basic introduction to the concept of symmetry and its use for simplifying equations. Chapter 5 discusses a variety of estimation and approximation techniques. Chapter 7 discusses how and when to generalize equations. Finally, Chapter 8 provides a few more instructive examples that illustrate several problem-solving techniques while reinforcing some of the concepts introduced in earlier chapters.

There are many interrelations and cross-references among the chapters. Generalization, which is discussed in Chapter 7, is closely related to the subject of special cases introduced in Chapter 3. Similarly, dimensional analysis and scaling (Chapter 6) are closely related to the study of units and dimensions (Chapter 1). Notice that these linked chapters are spaced apart in the book. This placement is intended to give the reader time to think about the material before it is revisited. Despite the interconnections among them, the chapters are, for the most part, self-contained and need not necessarily be read in numerical sequence. The book can be used to construct modules for a quick study of a more focused topic. Two such possible modules are:

(1) dimensional analysis and scaling: Chapters 1 and 6;
(2) special cases, approximation, and generalization: Chapters 3, 5, and 7.

We have also tried to enliven the discussion by identifying the nationality and approximate era of some of the luminaries who made major contributions to the scientific and mathematical literature that we have drawn upon, and by providing a few historical anecdotes. While this material is not necessary for dealing with the equations themselves, we hope that you find it entertaining and that it will provide some insight into how science often has progressed along an unexpected or a roundabout path.

If you have suggestions for additional material that could be included in future editions of this book, find errors or descriptions that are unclear, or wish to provide feedback of any other type, we would be happy to hear from you.

Matt A. Bernstein
Rochester, Minnesota
mbernstein@mayo.edu
William A. Friedman
Seattle, Washington
friedman@physics.wisc.edu

December 2008

ACKNOWLEDGEMENTS

Several individuals offered their time and effort to review the manuscript and provide comments. We especially want to thank Juliet Bernstein, who offered many insightful suggestions, which have greatly improved the clarity of the book. Kristi Welle reviewed several of the example problems, and Linda Greene proofread many of the pages. We thank Professor Željko Bajzer for showing us the curious integral that is discussed in Example 2.1. We thank them all for their outstanding efforts and emphasize that any remaining mistakes and inaccuracies are solely our own.

Individually, we would like to express our gratitude.

MAB: I thank my friends and coworkers at the Mayo Clinic and also in the broader magnetic resonance imaging community. It has been a privilege to work with such outstanding people during such an exciting time. I am also indebted to my entire family and especially to my wife Rhoda Lichy for her unwavering support and patience during the writing of this book.

WAF: I am grateful to the many students at all levels with whom I have had contact during my nearly 40 years of teaching, and to my late wife Cheryl who was by my side during that journey.

LIST OF WORKED-OUT EXAMPLE PROBLEMS

* Indicates a worked-out example problem that uses somewhat more advanced mathematical techniques compared to the others.

1 .

EQUATIONS REPRESENTING PHYSICAL QUANTITIES

Some equations model systems or processes that occur in the real, physical world. Most of the variables that appear in these equations have dimensions, and they carry certain physical units. For example, a variable d describing distance has the dimension of length and carries a specific unit such as meters, microns, or miles. The numerical value of the variable d is given as a multiple of the unit we choose, and the specific unit is usually chosen so that the numerical values are convenient to work with.

Without a unit, the physical meaning of the numerical value associated with a dimensioned variable contains no useful information. For example, to say the distance between points A and B is "$d = 8$" is not useful for scientific and engineering purposes. We also have to specify a unit of length, such as $d = 8\,\text{in}$, $d = 8\,\text{m}$, or $d = 8$ light-years, each of which describes very different quantities in the physical world.

A number that does not carry any physical units, e.g., 1, –2.23, or π, is said to be *dimensionless*. There are some dimensionless quantities that nonetheless can carry units. One well-known example is an angle θ. Angles are dimensionless because they represent the ratio of two lengths, namely the subtended arc length on a circle divided by the radius r of that circle. The natural unit for θ is the radian. The value of the angle for one complete circular revolution is 2π radians, which follows from the fact that the corresponding arc length is the circumference of the circle, or $2\pi r$. Alternatively, the degree is a commonly

Thinking About Equations: A Practical Guide for Developing Mathematical Intuition in the Physical Sciences and Engineering, by Matt A. Bernstein and William A. Friedman
Copyright © 2009 John Wiley & Sons, Inc.

used unit to measure angles. There are $360°$ in one complete circular revolution, so the conversion factor between radians and degrees is

$$1\,\text{rad} = 360°/2\pi \approx 57.3°. \tag{1.1}$$

We have a choice of which of these units to use.

At first, it might seem like keeping track of the units associated with each variable in an equation is an inconvenience, akin to carrying extra baggage. As explored in this chapter, however, the use of units in fact can help us to better understand equations that contain variables representing physical quantities. Keeping track of dimensions and units can also uncover errors and can simplify work. This theme recurs elsewhere in this book, especially in Chapter 6, where the topics of dimensional analysis and scaling are discussed.

Some units have long and interesting histories, which illustrate their importance in science, engineering, and commerce. In ancient times, balance scales were commonly used to measure weight. The unknown weight of an object was measured by counting the number of unit weights required to counterbalance it. The carob tree is grown in the Mediterranean region, and its fruit is a pod that contains multiple seeds. It was found that the weight of the carob seeds varied little from one to the next. Also, it was relatively easy to get a uniform set of seeds. The heavier or lighter seeds could be eliminated from the collection because their weight correlated well with their size.

So it became convenient to use a group of carob seeds of uniform size to counterbalance the unknown quantity on the other side of the scale. The weight of the carob seeds was also of a convenient magnitude for weighing small objects like gemstones. The relative weight of the unknown object was quite accurately expressed in terms of the equivalent number of carob seeds, and this practice became a standard for commerce. Measured in modern units, a typical carob seed has a mass of approximately $0.20\,\text{g}$. Today, the unit *carat* is used to measure the mass (or the equivalent weight) of gemstones. A carat is defined to be exactly $0.20\,\text{g}$, and its name is derived from the name of the carob tree and its seeds.

1.1 SYSTEMS OF UNITS

Many different systems of units have been devised. For most scientific and engineering work today, the preferred units are in the "SI" system. This designation comes from the French "Système International d'Unités" (International System of Units). SI units are based on quantities with the seven fundamental dimensions listed in Table 1.1. Note that three of these fundamental units, the meter, kilogram, and second, were carried over from the older MKS system of units for the quantities of length, mass, and time when the SI system was developed in 1960.

TABLE 1.1. The Seven Fundamental SI Units.

Dimension	SI Unit
Length	meter or metre (m)
Mass	kilogram (kg)
Time	second (s)
Electric current	ampere (A)
Thermodynamic temperature	kelvin (K)
Amount of substance	mole (mol)
Luminous intensity	candela (cd)

SI units have gained popularity for several reasons. First, they use prefixes (e.g., nano, milli, kilo, mega) based on powers of 10. Prefixes allow the introduction of related units that are appropriate over a wide range of scales. For example, the unit of $1\,nm$ is equal to $10^{-9}\,m$. The powers of 10 also make conversion relatively simple, for example, converting units of area:

$$1m^2 = \left(1 \times 10^9 \, nm\right)^2 = 10^{18}\,nm^2. \tag{1.2}$$

In contrast, the imperial (sometimes called "British") system of units contains conversion factors that are usually not integer powers of 10. For example, to express $1\,yd^2$ in terms of square inches, we have to calculate 36×36:

$$1yd^2 = (36\,in)^2 = 1296\,in^2. \tag{1.3}$$

Another advantage of the SI system is that it contains many named, derived units such as the watt to measure power. The addition of these derived units is one of the major changes between the MKS and SI systems. The derived units in the SI system are *coherent*, that is, each one can be expressed in terms of a product of the fundamental units (or other derived units) and a numerical multiplier that is equal to 1. For example, the SI unit of electrical charge is the coulomb, and

$$
\begin{aligned}
1\ \text{coulomb} &= 1\ \text{ampere} \times 1\ \text{second} = 1\ A \cdot s \\
1\ \text{coulomb} &= 1\ \text{farad} \times 1\ \text{volt} = 1\ F \cdot V \\
1\ \text{coulomb} &= 1\ \text{volt} \times 1\ \text{second} \div 1\ \text{ohm} = 1\ V \cdot s \cdot \Omega^{-1}.
\end{aligned} \tag{1.4}
$$

The derived SI unit for power is the watt, which is equal to 1 joule per second. On the other hand, a unit of power in the imperial system, 1 horsepower, equals to $550\,ft \cdot lb/s$. The simple conversion factors in the SI system also make it easy to decompose all of the derived units back into integer powers of the fundamental units. For example,

$$1 \text{ newton} = 1\,\text{m}\cdot\text{kg}\cdot\text{s}^{-2}(\text{force}),$$
$$1 \text{ watt} = 1\,\text{m}^2\cdot\text{kg}\cdot\text{s}^{-3}(\text{power}), \tag{1.5}$$
$$1 \text{ ohm} = 1\,\text{m}^2\cdot\text{kg}\cdot\text{s}^{-3}\cdot\text{A}^{-2}(\text{electrical resistance}).$$

As discussed in Chapter 6, the decomposition illustrated in Equation 1.5 is particularly useful for dimensional analysis.

1.2 CONVERSION OF UNITS

Scientists and engineers are trained to work with SI units. Inevitably, however, we encounter units from other dimensional systems that require conversion back and forth to the SI system. For example, the speed of a car is commonly expressed in miles per hour or kilometers per hour, but rarely in the SI unit of meter per second. Similarly, household electrical energy usage is billed in kilowatt-hours rather than the SI unit of joules. Sometimes, the scale of the SI unit is not very convenient. For example, a kilogram is a very large unit in which to express the mass of an individual molecule, and a meter is a very short unit for interstellar distances. Rather than relying solely on the power-of-10 prefixes mentioned previously, more convenient, non-SI units like the atomic mass unit (amu or u) or light-year are sometimes used. Fortunately, the conversion of units is straightforward, as illustrated in the following example.

EXAMPLE 1.1

Convert 1 mi/h to the SI unit for speed, meter per second. There are 5280 ft per mile, 12 in per foot, and 2.54 cm per inch.

ANSWER

Unit conversion is readily accomplished with multiplication by a string of conversions factors, each of which is dimensionless and equals to 1, such as $1 = (2.54\,\text{cm})/(1.00\,\text{in}) = 2.54\,\text{cm/in}$. We multiply together powers (or inverse powers) of the conversion factors so that all of the units "cancel," except for the desired result:

$$1\frac{\text{mi}}{\text{h}} = 1\frac{\text{mi}}{\text{h}} \times \frac{5280\,\text{ft}}{\text{mi}} \times \frac{12\,\text{in}}{\text{ft}} \times \frac{2.54\,\text{cm}}{\text{in}} \times \frac{1\,\text{m}}{100\,\text{cm}} \times \frac{1\,\text{h}}{60\,\text{min}} \times \frac{1\,\text{min}}{60\,\text{s}}. \tag{1.6}$$

Gathering the numerical terms,

$$1\,\text{mi/h} = \frac{12 \times 5280 \times 2.54}{100 \times 60 \times 60}\,\text{m/s} = 0.44704\,\text{m/s}. \tag{1.7}$$

As elementary as Example 1.1 seems, errors in unit conversion are not uncommon and can have disastrous consequences. A well-known example occurred on September 23, 1999, when an unmanned orbiting satellite approached Mars at too low an altitude and crashed into the red planet. A subsequent investigation by NASA revealed that engineers failed to properly convert the imperial system unit of force used to measure rocket thrust (the pound-force) into the SI unit force, the newton.

Another example of confusion caused by the improper conversion of units occurred over 300 years earlier in connection with Sir Isaac Newton's work on the theory of gravitation. In 1679, Newton consulted a sailor's manual to obtain numerical values that he used to check the predictions of his theory for the speed of the Moon as it orbits the Earth. That speed was known in Newton's time from the Moon's observed orbital period and its estimated distance from the Earth, which was deduced from the observation of eclipses. Newton, however, did not know that the term "mile" in the sailor's manual referred to a nautical mile, which is approximately 15% longer than the statute mile (5280 ft) with which he was familiar. This confusion led to a 15% discrepancy between his prediction for the speed of the Moon and the accepted value. Discouraged by this, Newton abandoned his correct approach and searched for an alternative theory. This detour delayed Newton's work on gravity by approximately 5 years. Eventually, of course, Newton discovered the error concerning units, and his theory of gravitation has become the basis for much of modern space flight.

1.3 DIMENSIONAL CHECKS AND THE USE OF SYMBOLIC PARAMETERS

Anytime we equate one term to another, they both must have the same dimensions for the expression to make physical sense. We cannot equate a term with the dimension of length to a term with the dimension of mass. Using the basic rules of algebra, we can extend this principle to say that whenever we add or subtract terms, they must also have the same dimensions. We say that such an expression is dimensionally consistent or dimensionally *homogeneous*. We can always add zero to or subtract zero from any equation. Whenever we do so, we will assume that the zero carries the appropriate dimensions.

If an equation is dimensionally inconsistent, we can recognize immediately that it must be flawed. The inconsistency might have arisen because the equation's construction was based on faulty principles, or because its derivation contained an algebraic error. The converse is not necessarily true. If an equation is dimensionally consistent, it does not mean that it is necessarily correct, only that it *could be* correct.

Consider the following alternative expressions both intended to describe the height y of a ball thrown in the air, as a function of time t,

$$y(t) = 3 + 2t - 4.9t^2 \tag{1.8}$$

and

$$y(t) = y_0 + v_{0y}t - \frac{1}{2}gt^2, \quad y_0 = 3\,\text{m}, v_{0y} = 2 \text{ m/s, and } g = 9.8\,\text{m/s}^2, \tag{1.9}$$

where t is measured in seconds. These two expressions might appear equivalent at first glance, and Equation 1.8 is more compact. Retaining symbols as in Equation 1.9, however, has several important advantages. First, a quick, visual check confirms that we are adding and subtracting terms that all have the same units, in this case meters. Equation 1.8 is not dimensionally homogeneous unless we assume that the units are implied, i.e., in the first term, "3" really means "3 m," and similarly for the other numerical coefficients. It is easy to apply this assumption inconsistently, leading to errors. On the other hand, retaining symbols and checking units can bring our attention to a typographic or careless error. For example, the incorrect exponent can be spotted easily

$$y(t) = y_0 + v_{0y}t - \frac{1}{2}gt^3 \quad [\text{incorrect}], \tag{1.10}$$

because the last term has the incorrect units of m·sec instead of meters. Common errors like these can be quite difficult to uncover when the numerical format of Equation 1.8 is used, where the dimensions of the numerical factors are implied, rather than given explicitly. The use of the symbolic format as in Equation 1.9 avoids many of these problems.

The symbolic format of Equation 1.9 also allows the acceleration and initial height and initial velocity for the trajectory of the ball to be easily extracted. The initial height is $y(t)|_{t=0} = y_0$, and differentiating $y(t)$ with respect to time yields the initial velocity and the acceleration:

$$\frac{dy}{dt}\bigg|_{t=0} = v_{0y}$$
$$\frac{d^2y}{dt^2} = -g \tag{1.11}$$

Equation 1.8 can also be differentiated with respect to time. With a proper choice of notation, however, symbols such as v_{0y} generally do a better job evoking the physical meaning of the parameters than the numerical values appearing in Equation 1.8. Usually, the more complicated the analytic expression, the greater the advantage of retaining the symbolic notation becomes.

If we assign numerical values to quantities appearing in equations, we also have to be careful about their units. For example, the expression (1 m + 1 cm) combines two quantities that have the dimension of length but are measured in different units. To reduce the expression correctly to 1.01 m or 101 cm (instead of "2") naturally requires proper unit conversion.

Finally, the basic expression for $y(t)$ in Equation 1.9 remains valid regardless of which system of units is chosen, which is not true for Equation 1.8. To convert the expression for $y(t)$ in Equation 1.9 into units of feet, we only need to convert the given parameters to a new set with the desired units. This is easy to do using the method illustrated in Example 1.1. Using Equation 1.9, the values of the coefficients become $y_0 = 9.84$ ft, $v_{0y} = 6.56$ ft/s, and $g = 32.2$ ft/s^2.

To summarize, the symbolic format of Equation 1.9 is preferred over the numerical format of Equation 1.8 because it facilitates dimensional checks, makes no assumptions about the dimensions of the parameters, and is easier to translate between systems of units. For computational work, the symbolic format of Equation 1.9 is also preferable to the "hard-coded" format of Equation 1.8, because it provides an easier and less error-prone way to pass the values of the coefficients between program modules such as subroutines.

1.4 ARGUMENTS OF TRANSCENDENTAL FUNCTIONS

Because added or subtracted terms must have the same dimensions, the following section will show that we can infer that the arguments of many transcendental functions are dimensionless. The trigonometric functions like sine, cosine, tangent, and secant are transcendental, as are the exponential functions, which also include hyperbolic sine and hyperbolic cosine. They are distinguished from algebraic functions, which include polynomials, square roots, and other simpler functions.

These functions are often expressed in terms of an infinite series expansion. Consider the well-known series expansion for the sine function:

$$\sin(u) = u - \frac{u^3}{6} + \frac{u^5}{120} - \frac{u^7}{5040} + \dots \tag{1.12}$$

Because the coefficients of 1/6, 1/120, etc., are dimensionless numbers, u cannot carry any physical units either. Suppose, for example, that u had the dimensions of power, measured in units of "watts." Because we cannot subtract a term with units of watts^3 from a term with units of watts, Equation 1.12 would not make physical sense. Therefore, the argument of the transcendental function $\sin(u)$ must always be dimensionless.

Typically, the arguments of trigonometric functions are angles. As mentioned earlier, angles are dimensionless, and their SI unit is the radian. It is important to remember that many common mathematical formulas involving trigonometric functions such as the series expansion

$$\cos\theta = 1 - \frac{\theta^2}{2} + \frac{\theta^4}{24} - \frac{\theta^6}{720} + \dots \tag{1.13}$$

or the derivative

$$\frac{d}{d\theta}\sin\theta = \cos\theta \tag{1.14}$$

are not valid unless the angle θ is measured in radians. For example, Equation 1.12 implies that the sine of a small angle ($\ll 1\,\text{rad}$) is approximately equal to the angle itself, $\sin\theta \approx \theta$. With the use of a scientific calculator, the reader can easily verify that, to five significant figures, $\sin(0.01\,\text{rad}) = 0.01000$. On the other hand, $\sin(0.01\,^\circ) = 0.00017453$, which reflects the conversion factor ($2\pi/360 \approx 0.01745$) stated in Equation 1.1.

Next, consider an exponential function and its series expansion:

$$f(t) = e^{3t} = 1 + 3t + 4.5t^2 \dots \tag{1.15}$$

Equation 1.15 is dimensionally consistent provided that t is a dimensionless variable. If, however, the variable t represents time measured in seconds, then Equation 1.15 does not make sense unless we assume that the units are implied, i.e., "3" really means "$3\,\text{s}^{-1}$" and "4.5" really means "$4.5\,\text{s}^{-2}$." The discussion following Equation 1.8 showed how this type of assumption can lead to problems. Instead, it is better to write

$$f(t) = e^{\lambda t} = 1 + \lambda t + \frac{(\lambda t)^2}{2} + \dots \tag{1.16}$$

with $\lambda = 3\,\text{s}^{-1}$. The argument of the transcendental function in Equation 1.16 is now the product λt, which is dimensionless, as is the third term in the expansion, $\tfrac{1}{2}(\lambda t)^2$.

Among the transcendental functions, the logarithm provides an interesting special case. Consider $\log(x/a)$, where the ratio (x/a) is dimensionless. For example, both x and a might have the dimension of length, measured in units of meters. Suppose that $x = 3\,\text{m}$ and $a = 2\,\text{m}$. Logarithms reduce the operation of division to subtraction, i.e.,

$$\log\left(\frac{x}{a}\right) = \log\left(\frac{3\,\text{m}}{2\,\text{m}}\right) = \log\left(\frac{3}{2}\right) = \log(3) - \log(2). \tag{1.17}$$

All of the operations in Equation 1.17 are valid, and note that Equation 1.17 holds regardless of the logarithm's base, e.g., 10, 2, or e.

We might encounter a symbolic expression containing a term $\log(x)$, where x is not dimensionless. We cannot immediately conclude that the *entire* expression is incorrect. The expression might also include another term of the form $-\log(a)$, or equivalently $+\log(1/a)$, where a has the same units as x. Then, even though the variable x appearing in $\log(x)$ is dimensioned, the entire expression can be correct.

1.5 DIMENSIONAL CHECKS TO GENERALIZE EQUATIONS

The use of dimensional checks allows us to generalize equations and even generate new ones. Suppose we use integration by parts or a table of integrals and find

$$F(x) = \int xe^x dx = e^x(x-1) + C, \qquad (1.18)$$

where C is a constant. We know that the variable x in Equation 1.18 must be dimensionless, because it is the argument of the exponential function. Using dimensional checks, we can generalize Equation 1.18 to evaluate

$$G(x,a) = \int xe^{ax} dx \qquad (1.19)$$

without the need for additional integration. The product (ax) must be dimensionless because it appears as the argument of the exponential Equation 1.19. We are free to assume that the variable x has dimensions of length, measured in the unit of meters, and we will do so. In that case, a must have units of m^{-1}. So the task is to use dimensional checks to place the appropriate power of a into each term of the right-hand side of Equation 1.18. Recognizing that each of the terms of $G(x,a)$ must have units of meters squared (because the differential dx carries the same units as x, see exercise 1.12), we can infer that

$$G(x,a) = \int xe^{ax} dx = \frac{e^{ax}}{a}\left(x - \frac{1}{a}\right) + C'. \qquad (1.20)$$

Along with the dimensional check, we also used the fact that $G(x, a)$ reduces to $F(x)$ when $a = 1\,m^{-1}$. Naturally, for a further check, we can differentiate the right-hand side of Equation 1.20. Although we do not know the values of the integration constants C and C', we do know their units. C is dimensionless, and C' carries the same units as x/a, namely meters squared.

Equation 1.20, which was generated with the aid of a dimensional check, can be extended to evaluate related integrals, up to the additive constant (see exercises 1.2 and 1.13). The partial derivative of $G(x, a)$ with respect to a yields

$$\frac{\partial G(x,a)}{\partial a} = \int x \left(\frac{\partial e^{ax}}{\partial a} \right) dx = \int x^2 e^{ax} dx$$

$$= \frac{\partial}{\partial a} \left[\frac{e^{ax}}{a} \left(x - \frac{1}{a} \right) + C' \right] = \frac{e^{ax}}{a} \left(x^2 - \frac{2x}{a} + \frac{2}{a^2} \right) \tag{1.21}$$

Equation 1.21 implies that

$$\int x^2 e^{ax} dx = \frac{e^{ax}}{a} \left(x^2 - \frac{2x}{a} + \frac{2}{a^2} \right) + C''. \tag{1.22}$$

Starting with Equation 1.18, the use of dimensional checks followed by partial differentiation yielded a new expression given in Equation 1.22. This sequence of operations can be quite useful for evaluating a variety of indefinite and definite integrals, but we always have to be very careful to follow the rules of calculus. In the example of Equation 1.21, it would *not* be valid to differentiate with respect to x, because x is the integration variable.

1.6 OTHER TYPES OF UNITS

Up to this point, Chapter 1 has focused on variables that carry physical units that describe length, mass, electrical charge, etc. This allowed us to draw useful inferences about equations composed of these variables. There are many other types of units, however, that are not associated with the physical sciences. One example is a monetary unit, like a dollar or a euro. Another example is a unit like the number of soldiers per battalion, which might be used in a logistical calculation to find the required amount of rations for a month.

Although units like $ or battalion^{-1} are not part of the SI system, they are often very convenient (for example, see exercise 1.14). Equations containing variables that carry these types of units still must be dimensionally homogeneous, provided that the system of units is applied in a consistent manner. Thus, the dimensional checks and unit conversion methods introduced previously in this chapter apply.

Some equations seem to model the physical world quite well but appear to be dimensionally inconsistent. For example, we might find that the time t it takes to finish a task in the office is well described by the equation "$t = 20\,\text{min} + \text{three times the number of phone call interruptions received}$." This equation seems to be dimensionally inconsistent, because an apparently dimensionless quantity (three times the number of phone calls) is added to a dimensioned quantity (20 min). Whenever this type of expression accurately models the physical world, however, there is an implied conversion factor. In this case, the implied conversion factor is 3 min/phone call.

1.7 SIMPLIFYING INTERMEDIATE CALCULATIONS

Calculations often require many intermediate steps to obtain the desired result. Sometimes when performing calculations with symbolic variables, it is convenient to *temporarily* choose a dimensional system (i.e., a set of units) so that the numerical values of some of the physical constants are equal to 1. This trick can simplify algebraic manipulation, whether it is performed with paper and a pencil or with symbolic manipulation software. Symbolic coefficients, such as those appearing in Equation 1.9, can also temporarily be set equal to 1. After the calculation is completed, the symbols are replaced to make the result dimensionally consistent, as was done to derive Equation 1.20. The entire procedure is illustrated in Example 1.2.

Setting quantities equal to 1, even temporarily, seems like it could lead to incorrect or confusing results. Trouble can be avoided, however, by following these two rules:

Rule 1: Never set a dimensionless quantity equal to 1.

Violating this rule clearly can lead to logical inconsistencies. For example, if we set the dimensionless number 2 equal to 1, we can immediately write an incorrect equation such as "1 + 1 = 4." To perform rough *estimates* (as opposed to exact calculations), we sometimes neglect factors of 2, 4, π, and so on. Estimation is discussed in Chapter 5 and is not our focus here.

Rule 2: When setting a collection of dimensioned quantities equal to 1, never choose this group to be sufficiently large so that a dimensionless product can be formed from them. To do so would result in a dimensionless quantity, i.e., that product, being set equal to 1 in violation of rule 1.

The meaning of rule 2 is illustrated by the following case. Suppose we temporarily set each of three dimensioned quantities q_1, q_2, and q_3 equal to 1 and then we find any exponents a, b, and c such that the product $q_1^a \times q_2^b \times q_3^c$ is dimensionless. Then, rule 2 says that we have gone too far. We need to restore at least one of the q's back to its original value. Rule 2 also implies that we should never simultaneously set two different quantities of the same dimension equal to 1, for example, the height and width of a rectangle. Rule 2 ensures that there is only a single way to return the dimensioned variables back into the final expression when making it (explicitly) dimensionally consistent. This uniqueness property is further explored in exercise 1.9.

EXAMPLE 1.2

To illustrate why it is convenient to temporarily set constants equal to 1, consider the example of Compton scattering, named in honor of the American physicist Arthur H. Compton who published a paper on this effect in 1923. Compton scattering describes the scattering of an X-ray photon from a free electron, which is assumed to initially be at rest. The X-ray photon loses some of its energy to the electron, which results in a reduction of the frequency of the scattered X-ray. The amount of energy the X-ray photon loses (and therefore the frequency of the scattered radiation) depends on the angle through which it is scattered. If it is scattered straight back in the direction from which it came (i.e., $\phi = 180°$), then it loses the maximum possible amount of energy. Compton scattering is important for many applications, including medical imaging methods that use X-rays. One such method is computed tomography. The concepts from modern physics that are applied to set up the equations are described in detail in many physics books, and we only give a brief outline here.

In this example, we will solve the equations for Compton scattering with two procedures: (a) using standard units and (b) using a dimensional system where Planck's constant and the speed of light each equal to 1. Part (b) of this example illustrates the algebraic simplification that can result.

ANSWER

(a) When the X-ray photon scatters from the electron, conservation of energy yields the relation

$$hf_0 + Mc^2 = hf + E, \tag{1.23}$$

where h is Planck's constant, f_0 is the frequency of the incident X-ray, M is the rest mass of the electron, c is the speed of light, f is the frequency of the scattered X-ray, and E is the final, total energy of the electron. There is also an identity from the special theory of relativity that relates the final energy, mass, and momentum P of the electron:

$$E^2 = \left(Mc^2\right)^2 + (Pc)^2. \tag{1.24}$$

Equations 1.23 and 1.24 are both dimensionally consistent. The magnitudes of the initial and final values of the momentum of the X-ray are given by (hf_0/c) and (hf/c), respectively. The X-ray is scattered through an angle ϕ, so that the conservation of momentum and the law of cosines imply

$$P^2 = \left(\frac{hf_0}{c}\right)^2 + \left(\frac{hf}{c}\right)^2 - 2\left(\frac{h^2 f_0 f}{c^2}\right)\cos\phi. \tag{1.25}$$

The final energy E and momentum P of the electron can be eliminated from Equations 1.23–1.25. After some algebra (see exercise 1.10), the desired result for the final frequency of the X-ray is obtained:

$$f = \frac{f_0}{1 + \dfrac{hf_0}{Mc^2}(1 - \cos\phi)}. \tag{1.26}$$

(b) The appearance of Equations 1.23–1.25 is simplified if we temporarily set Planck's constant h and the speed of light c both equal to 1. (This is clearly not true in SI units, in which $h = 6.626068 \times 10^{-34}$ J·s and $c = 2.9979 \times 10^8$ m/s.) The simplified versions of Equations 1.23–1.25 become

$$\begin{aligned} f_0 + M &= f + E, \\ E^2 &= M^2 + P^2, \\ P^2 &= f_0^2 + f^2 - 2f_0 f \cos\phi. \end{aligned} \tag{1.27}$$

The algebraic manipulation is now considerably simpler, because we do not have to carry the factors of h and c through each step. After the simplified algebra, we find that

$$"f" = \frac{f_0}{1 + \dfrac{f_0}{M}(1 - \cos\phi)} \tag{1.28}$$

We know that Equation 1.28 cannot be the final, correct result for f, because it is not dimensionally consistent; the dimensioned term $(f_0/M) \times (1 - \cos\phi)$ is added to the dimensionless quantity 1 in the denominator. We must properly reinsert the dimensioned quantities into the final expression to get the correct answer.

To replace the correct factors of h and c to Equation 1.28, it is convenient to decompose each of the dimensions of each of the variables into fundamental SI units. It is conventional to denote the units of a variable by placing square brackets around it. For the variables in this example,

$$\begin{aligned} [f_0] &= [f] = s^{-1}, \\ [M] &= kg, \\ [\phi] &= rad, \\ [c] &= m \cdot s^{-1}, \\ [h] &= J \cdot s = m^2 \cdot kg \cdot s^{-1}. \end{aligned} \tag{1.29}$$

Naturally, $[f_0/M] = [f_0]/[M]$, so that the ratio (f_0/M) has units $\text{kg}^{-1}\cdot\text{s}^{-1}$. We now reintroduce the powers of h and c needed for Equation 1.28 to be dimensionally consistent. The result is Equation 1.26. This can be seen by inspection, but we can also reach the same conclusion more methodically, as described next.

Because f_0 and f both have the same dimensions, all that is required to make Equation 1.28 dimensionally consistent is to replace the powers of h and c so that the term (f_0/M) in the denominator becomes dimensionless. We seek values of the exponents a and b such that

$$\left[h^a c^b f_0 M^{-1}\right] = 1 = 1\text{m}^0 \times 1\text{s}^0 \times 1\text{kg}^0, \tag{1.30}$$

where we omitted the other four fundamental units (like amperes and candelas) because they do not enter into this particular equation. From Equations 1.29 and 1.30, we obtain

$$\left[h^a c^b f_0 M^{-1}\right] = \left(\text{m}^2\cdot\text{kg}\cdot\text{s}^{-1}\right)^a \left(\text{m}\cdot\text{s}^{-1}\right)^b \left(\text{s}^{-1}\right)\left(\text{kg}^{-1}\right) = \text{m}^{2a+b}\text{s}^{-a-b-1}\text{kg}^{a-1}. \tag{1.31}$$

Equating the exponents of each factor in Equations 1.30 and 1.31 yields a set of linear equations:

$$\begin{aligned} 2a+b &= 0, \\ -a-b-1 &= 0, \\ a-1 &= 0. \end{aligned} \tag{1.32}$$

These can be readily solved to give $a = 1$ and $b = -2$, which agree with Equation 1.26.

In Example 1.2, temporarily choosing a dimensional system so that $c = 1$ and $h = 1$ has allowed us to get the correct answer while simplifying the algebra. Note that from Equation 1.29, simultaneously setting both $c = 1$ and $h = 1$ did not violate rule 2, because no power of c can cancel the basic SI unit kilogram appearing in h to form a dimensionless quantity.

Although the procedure illustrated in Example 1.2b can simplify algebraic calculation, it is not to everyone's liking. Even if the two stated rules are followed carefully, the ability to perform dimensional checks on each intermediate step of the calculation is lost. Instead, many prefer to solve the problem using standard units as in Example 1.2a but with variable substitutions such as

$$\begin{aligned} u &= Mc^2, \\ v &= hf_0, \\ w &= hf. \end{aligned} \tag{1.33}$$

This method allows for algebraic simplification and still permits dimensional checks at each step. One drawback of the substitutions of Equation

1.33 is that the constants can reappear in the expression when we calculate derivatives or integrals. For example, the constant c reappears if we need to differentiate some function y with respect to M:

$$\frac{dy}{dM} = c^2 \frac{dy}{du}. \qquad (1.34)$$

The choice of which, if any, of these simplification methods to use depends on the complexity of the specific problem and is mainly a matter of personal preference.

EXERCISES

(1.1) Convert 1 light-year into meters. Use the speed of light $c = 2.9979 \times 10^8$ m/s.

(1.2) Suppose that

$$\int x^3 \sin x dx = (3x^2 - 6)\sin x + (6x - x^3)\cos x + C.$$

(a) Without further integration, use dimensional checks to evaluate $\int x^3 \sin(ax)dx$. Assume that x has the dimensions of length.

(b) Use the resulting expression to evaluate $\int x^4 \cos(ax)dx$ by partial differentiation.

(1.3) A dimensionless variable u is raised to the power b. Show that the exponent b must also be dimensionless. Hint: use $u^b = \exp(\ln u^b) = \exp(b \ln u)$.

(1.4) Suppose that $g = 9.8$ m/s^2, $v = 50$ m/s, $h = 12$ m, $\omega = 30$ rad/s, $\theta = 2.5$ rad, and t is time measured in seconds. Identify which of the following equations cannot be valid based on a dimensional check:

(a) $v = ht$;

(b) $v = \sqrt{2gh}$;

(c) $\theta = \sin \omega t + \dfrac{h}{vt}$;

(d) $h = vt + \exp(-gt)$.

(1.5) Each of the following mathematical expressions has one term that is dimensionally inconsistent with the other two. Find the inconsistent term, and correct it by inserting the appropriate power of a. Assume x is a length:

(a) $a^2 e^{ax} + \dfrac{9\sin(ax)}{x^2} - 3(ax)^5$;

(b) $\dfrac{\sqrt{ax}}{2+(ax)^2} + \dfrac{4\sin ax}{x} + \exp\left[-\dfrac{1}{2}(ax)^2\right]$;

(c) $x\arccos\left(a^{0.25}\sqrt{x}\right) - x\sin\left(ax^2\right) + 3$.

(1.6) Look up the definitions of each of these derived SI units: (a) henry, (b) gray, and (c) lux. For each unit, describe what physical quantity it measures and then express the unit as a monomial containing powers of the seven fundamental units (i.e., see Equation 1.5). (A "monomial" is a polynomial with only one term.)

(1.7) Consider the quadratic equation $ax^2 + bx + c = 0$. Assume that x has the dimensions of length and that a is dimensionless. What are the dimensions of b and c? Verify that the solution

$$x = \frac{-b \pm \sqrt{b^2 - 4ac}}{2a}$$

is dimensionally consistent. If instead b is dimensionless, then what are the dimensions of a and c? In that case, is the solution for x still dimensionally consistent?

(1.8) Suppose a particle of mass m and kinetic energy E collides with another particle of mass m, which is initially at rest. We analyze the problem using the special theory of relativity, and to simplify the resulting equations, we temporarily set $m = 1$ and $c = 1$, where c is the speed of light. After some algebra, we find that the kinetic energy in the center-of-mass reference frame is given by

$$E_{cm} = 2\left(\sqrt{1 + \frac{E}{2}} - 1\right).$$

Reintroduce factors of m and c so that the expression is (explicitly) dimensionally consistent.

(1.9) To reduce the algebraic complexity of a problem, we set three dimensioned physical constants q_1, q_2, and q_3 each equal to 1, similar to Example 1.2. After completing the algebra, we then go on to replace powers of q_1, q_2, and q_3 so that the final expression is dimensionally consistent. We find that for one of the terms in our expression, we can replace the physical constants in two distinct ways: $\sqrt{q_1} \times q_2 \times q_3^2$ and $q_1^{3/2} \times \sqrt{q_2} \times q_3$ both make the expression dimensionally consistent. What went wrong? Hint: show that rule 2 was violated. (By this same method, show that, in general, whenever the replacement of the physical constants is not unique, rule 2 must be violated. This implies that if rule 2 is satisfied, then the replacement is unique.)

(1.10) Derive Equation 1.26 from Equations 1.23–1.25. Start by isolating E in Equation 1.23 and then squaring the result. Then derive Equation

1.28 from Equation 1.27. Do you find that the simplification justifies the potential pitfalls of setting the physical constants equal to 1?

(1.11) Interpret each of the following equations from a dimensional perspective. For example, the Pythagorean theorem $c^2 = a^2 + b^2$ is dimensionally consistent only if a, b, and c all have the same units, such as length:

(a) formula for an ellipse: $\left(\dfrac{x - x_0}{a}\right)^2 + \left(\dfrac{y - y_0}{b}\right)^2 = 1$;

(b) derivative of a function raised to a power: $\dfrac{d}{dx}(u^n) = nu^{n-1}\dfrac{du}{dx}$.

(1.12) Show that the differential dx must have the same units as the variable x. Hint: consider the integral $\int dx$.

(1.13) Suppose we know that $\displaystyle\int_{-\infty}^{\infty} \exp(-x^2)\,dx = \sqrt{\pi}$.

(a) Use a dimensional check to evaluate $\displaystyle\int_{-\infty}^{\infty} \exp(-ax^2)\,dx$.

(b) Use partial differentiation to evaluate $\displaystyle\int_{-\infty}^{\infty} x^4 \exp(-x^2)\,dx$.

(1.14) Suppose that four painters can paint two houses in 2 days. How many days does it take for three painters to paint five houses? Hint: form an expression for the house-painting rate r_p, measured in non-SI units:

$$r_p = \frac{2\ \text{houses}}{4\ \text{painters} \times 2\ \text{days}} = 0.25\ \text{house} \cdot \text{painter}^{-1}\text{day}^{-1}.$$

FURTHER READING

http://physics.nist.gov/cuu/Units/index.html

The web pages located in this URL from the National Institute of Standards and Technology contain many helpful links and much useful information about systems of units.

Szirtes T. 1997. *Applied Dimensional Analysis and Scaling*. New York: McGraw-Hill.

This book gives a very thorough discussion about dimensional consistency and systems of units. Among many other topics, Szirtes provides an interesting discussion regarding the dimensions of "dimensionless" quantities, which the reader might find useful for interpreting Equation 1.30.

2

A FEW PITFALLS AND A FEW USEFUL TRICKS

In this chapter, we discuss some pitfalls that frequently lead to errors when dealing with equations. We also present a few especially useful tricks that can simplify solving and working with equations. The material selected for this chapter represents only a small sample of the potential topics. Some of the pitfalls were chosen to illustrate themes that are repeatedly discussed in the book, while some of the tricks are used in example problems presented in subsequent chapters.

2.1 A FEW INSTRUCTIVE PITFALLS

The first example illustrates a common pitfall: the misapplication of a mathematical theorem. Theorems and their proofs are the stock and trade of mathematicians. In science and engineering, the proper application of these theorems is essential for performing many practical calculations.

Proper application of a theorem does not necessarily require one to prove it, or even to understand all of the logical steps in the proof. What is essential, however, is that all of the conditions stated in the theorem are satisfied. To ensure this requires that the conditions be understood. When all of the conditions of a theorem are not satisfied, incorrect numerical answers can result, as illustrated in the next example (also see exercise 3.6 in Chapter 3).

Thinking About Equations: A Practical Guide for Developing Mathematical Intuition in the Physical Sciences and Engineering, by Matt A. Bernstein and William A. Friedman
Copyright © 2009 John Wiley & Sons, Inc.

EXAMPLE 2.1

Consider the definite integral

$$I = \int_0^{2\pi} \frac{2dx}{5 + 3\cos x}. \tag{2.1}$$

At first glance, the integral in Equation 2.1 appears to be routine. We could always evaluate Equation 2.1 with numerical integration, but typically, we would apply the fundamental theorem of integral calculus instead. This theorem requires that we find an antiderivative $F(x)$ of the integrand $f(x) = 2/(5 + 3\cos x)$ and then evaluate it at the integration end points:

$$I = F(2\pi) - F(0). \tag{2.2}$$

As described in standard texts on calculus, the conditions of the fundamental theorem of integral calculus require that (1) the function $f(x)$ is continuous over the entire range of integration $[0, 2\pi]$, and (2) $F(x)$ is an antiderivative of f over the entire interval.

(a) Determine the antiderivative $F(x)$ of $f(x)$, in preparation for evaluating $F(x)$ at the integration end points. Specifically, show that the antiderivative of the integrand $f(x) = 2/(5 + 3\cos x)$ is

$$F(x) = \arctan\left(\frac{1}{2}\tan\left(\frac{x}{2}\right)\right) + C, \tag{2.3}$$

where C is any constant. Then, calculate the difference given in Equation 2.2.

(b) The integrand $f(x) = 2/(5 + 3\cos x)$ is continuous, and its values are always positive. Consequently, we expect the value of the integral I in Equation 2.1 to be a positive number. Compare the answer obtained in part (a) against this expectation.

ANSWER

(a) Recall that

$$\frac{d(\arctan u)}{du} = \frac{1}{1 + u^2} \tag{2.4}$$

and

$$\frac{d(\tan u)}{du} = \sec^2 u. \tag{2.5}$$

Applying Equations 2.4 and 2.5, and the chain rule to calculate the derivative of $F(x)$ yields

$$\frac{dF}{dx} = \frac{d}{dx}\left[\arctan\left(\frac{1}{2}\tan\left(\frac{x}{2}\right)\right)\right] = \frac{\frac{1}{4}\sec^2\left(\frac{x}{2}\right)}{1+\frac{1}{4}\tan^2\left(\frac{x}{2}\right)}. \tag{2.6}$$

To simplify Equation 2.6, we first divide the numerator and denominator by $\sec^2 x$. Then, we apply the trigonometric identities

$$\sin^2 u + \cos^2 u = 1 \tag{2.7}$$

and

$$\cos^2\left(\frac{u}{2}\right) = \frac{1}{2}(\cos u + 1) \tag{2.8}$$

to obtain

$$\frac{dF}{dx} = \frac{1}{4\cos^2\left(\frac{x}{2}\right)+\sin^2\left(\frac{x}{2}\right)} = \frac{1}{3\cos^2\left(\frac{x}{2}\right)+1} = \frac{2}{3\cos x + 5}. \tag{2.9}$$

From Equation 2.9, we conclude that the function $F(x) = \arctan(\tan(x/2)/2) + C$ is indeed an antiderivative of $f(x) = 2/(5 + 3\cos x)$.

The antiderivative $F(x)$ given in Equation 2.3 contains an arctangent, which is well known to have multiple branches. We will use the principal branch of the arctangent function

$$-\frac{\pi}{2} < \arctan x < \frac{\pi}{2}. \tag{2.10}$$

The convention for evaluating the arctangent with Equation 2.10 is almost universally adopted, as can be verified by placing any scientific calculator into "radian mode" and calculating various values of the arctangent function. Following that convention and applying the fundamental theorem to evaluate the integral in Equation 2.1, we obtain

$$I = \int_0^{2\pi} \frac{2dx}{5+3\cos x} = F(2\pi) - F(0) = 0 \quad [\text{incorrect}]. \tag{2.11}$$

The value $I = 0$ is the wrong answer! The evaluation of Equation 2.3 by numerical integration yields $I = 3.14159$ to six significant digits, which is an approximation to the correct answer, $I = \pi$.

(b) First, we check the algebraic steps performed in part (a). The division of the numerator and denominator of Equation 2.6 by $\sec^2 x$ is valid because $\sec^2 x$ is never zero. That operation of division is not the source of the error.

The plots of the functions $f(x)$ and its antiderivative $F(x)$ in Fig. 2.1 provide a clue about what went wrong. The point $x = \pi$ warrants closer inspection, because $F(x)$ is not continuous at $x = \pi$. The point $u = \pi/2$ lies at the boundary between two branches of the function $\tan(u)$. The point $u = \pi/2$ is also a singularity because $\tan u$ approaches $\pm\infty$ as u approaches $\pi/2$, with the "+" sign corresponding to the principal branch described in Equation 2.10. Although the integrand $f(x)$ is continuous and well behaved, the antiderivative $F(x)$ is discontinuous at $x = \pi$. Therefore, the derivative of $F(x)$ does not exist at $x = \pi$, and condition 2 required by the fundamental theorem of integral calculus is violated.

The second condition of the fundamental theorem of integral calculus states that $F(x)$ must be an antiderivative of f over the *entire* interval. Notice that this condition places additional requirements on $F(x)$, namely that it is also continuous and differentiable over the entire integration interval. The violation of the second condition at that single point $x = \pi$ by $F(x)$ is sufficient to invalidate Equation 2.11.

In the study of calculus and numerical analysis, we often encounter the useful adage that "integration is a smoothing process." For example, the plot of the absolute value function

$$f(x) = |x| \qquad (2.12)$$

is a v-shaped curve with a cusp, i.e., a discontinuous first derivative at $x = 0$. The antiderivative is

$$F(x) = \int |x| dx = \frac{1}{2}|x|x + C. \qquad (2.13)$$

Equation 2.13 represents a smoother, s-shaped curve with no cusp. Because integration is a smoothing process, the result of Example 2.1 is particularly surprising. There, the integrand $f(x)$ is continuous, yet the function $F(x)$ that we calculated to be its antiderivative has a discontinuity, so it is certainly less smooth.

The tangent function has multiple branches, that is,

$$\tan(u + \pi n) = \tan(u), \quad n = 0, \pm 1, \pm 2, \dots. \qquad (2.14)$$

According to its standard definition, a mathematical function must provide only a single output for each input. So, from Equation 2.14, we have to be careful when defining the arctangent function. According to Equation 2.14, the inverse of the tangent is multivalued and defined only up to the addition of an integer multiple of π, i.e., modulo π. A multivalued relation like this is not a function. In an attempt to avoid this problem, we carefully specified the use of the principal branch of the arctangent function described in Equation 2.10. Despite these good intentions, the integration path crossed the singularity of the function $\tan(x/2)$ at $x = \pi$, which was responsible for introducing the discontinuity in $F(x)$.

To sidestep this difficulty, we can construct a continuous function $F_c(x)$

$$F_c(x) = \begin{cases} \arctan\left(\dfrac{1}{2}\tan\left(\dfrac{x}{2}\right)\right), & 0 \le x < \pi \\[2mm] \dfrac{\pi}{2}, & x = \pi \\[2mm] \pi + \arctan\left(\dfrac{1}{2}\tan\left(\dfrac{x}{2}\right)\right), & \pi < x \le 2\pi \end{cases} \qquad (2.15)$$

which *is* an antiderivative of $f(x)$ over the *entire* interval $0 \le x \le 2\pi$. The continuous function $F_c(x)$ is plotted as the dotted line in Fig. 2.1. The function $F_c(x)$ satisfies the conditions required by the fundamental theorem of integral calculus. We then obtain

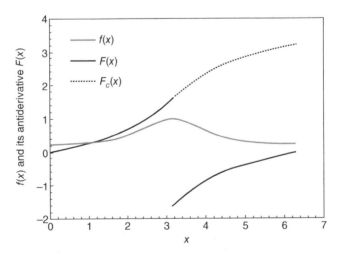

FIG. 2.1. The function $f(x)$ is continuous over the interval $0 \le x \le 2\pi$. The antiderivative $F(x)$ contains a discontinuity at $x = \pi$, which leads to a pitfall when applying the fundamental theorem of integral calculus. The problem is remedied by constructing an antiderivative function $F_c(x)$ that is continuous over the entire integration interval.

$$I = \int_{0}^{2\pi} \frac{2dx}{5+3\cos x} = F_c(2\pi) - F_c(0) = \pi, \qquad (2.16)$$

which is the correct numerical answer.

The second pitfall we discuss deals with adding (or subtracting) angles. This topic is related to Example 2.1 because the multivalued property of angles suggested by Equation 2.14 plays a central role.

When averaging a set of N quantities $(x_1, x_2, x_3, \ldots, x_N)$, we calculate the arithmetic mean, denoted by $\bar{x}$ or $\langle x \rangle$, with the standard formula

$$\langle x \rangle = \frac{1}{N} \sum_{i=1}^{N} x_i. \qquad (2.17)$$

Equation 2.17 is straightforward and might seem like an improbable source of a pitfall. If the quantities x represent angular measures, however, we can encounter problems.

Applying an integer number of complete circular revolutions to any angle yields an indistinguishable angle. Therefore, the measure of any angle θ can only be specified modulo $360°$ or 2π radians, i.e., θ is equivalent to $\theta + 2\pi n$, where $n = \pm 1, \pm 2, \ldots$. This complication is analogous to the multiple branches of the arctangent function discussed in connection with Equation 2.14 and causes ambiguity when calculating the average value of a set of angles.

Consider an example with $N = 2$ illustrated in Fig. 2.2. The angles $\theta_1 = 170°$ and $\theta_2 = -170°$. Applying Equation 2.17, we obtain

$$\langle \theta \rangle = \frac{170° - 170°}{2} = 0. \qquad (2.18)$$

Referring to Fig. 2.2, we can see that Equation 2.18 is not the expected answer, which is instead $\langle \theta \rangle = 180°$ or π radians.

How can we avoid this problem? One approach is to use the modulo $360°$ property of angles and then to manually insert the appropriate number of multiples of $360°$ based on the examination of Fig. 2.2. We could then modify Equation 2.18 as follows:

$$\langle \theta \rangle = \frac{170° + (-170° + 360°)}{2} = \frac{170° + 190°}{2} = 180°. \qquad (2.19)$$

Equation 2.19 now gives the desired answer but requires some manual intervention to construct. This method is not practical if we need to average thousands or millions of angles.

We can avoid these difficulties by averaging vectors instead of forming the algebraic sum of angles. We can express a vector in polar coordinates, i.e., in

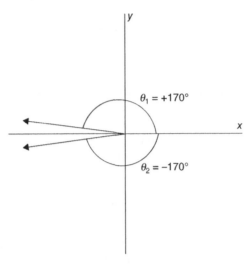

FIG. 2.2. Two angles are given by $\theta_1 = 170°$ and $\theta_2 = -170°$. The algebraic sum of those quantities is zero, so their arithmetic mean is zero. From the figure, however, we expect the average of the two angles to be $\langle\theta\rangle = 180°$, not zero. By averaging vectors, phasors, or complex exponentials instead of angular measures, this pitfall is avoided.

terms of its magnitude and the angle θ it makes with a fixed reference such as the x-axis. A vector in this representation is called a phase vector or *phasor*. To resolve our problem with averaging angles, we can construct a phasor of unit length $(\cos\theta, \sin\theta)$ corresponding to each angle. We then extract the value of $\langle\theta\rangle$ by examining the phase of the *vector sum* of all the individual unit phasors.

We can construct a diagram to find the vector sum of the individual unit phasors, but again, this is not very practical if we are averaging many angles. Instead, we can decompose the k-th unit phasor $\hat{p}_k$ into its components in the xy-plane, that is,

$$p_{kx} = \cos\theta_k,$$
$$p_{ky} = \sin\theta_k. \tag{2.20}$$

Notice that the phasor is unchanged if we add an integer multiple of 2π to θ_k because

$$\cos(\theta_k + 2\pi n) = \cos(\theta_k),$$
$$\sin(\theta_k + 2\pi n) = \sin(\theta_k). \tag{2.21}$$

Therefore, averaging unit phasors by using their Cartesian components sidesteps the ambiguity associated with Equations 2.18 and 2.19. We find

$$\langle p_x \rangle = \frac{1}{N} \sum_{k=1}^{N} \cos \theta_k \tag{2.22}$$

and

$$\langle p_y \rangle = \frac{1}{N} \sum_{k=1}^{N} \sin \theta_k. \tag{2.23}$$

The summations in Equations 2.22 and 2.23 provide numbers corresponding to the two orthogonal components resulting from the vector sum of the unit phasors. These two numbers determine the orientation corresponding to the average angle. To extract the numerical value of $\langle \theta \rangle$, we can use a four-quadrant arctangent function "ATAN2," which is provided as a built-in function in popular programming languages including C, FORTRAN, Java, and others. The ATAN2 function accepts two arguments, and based on their signs, it chooses the quadrant of the output value. ATAN2 returns an angular value in the interval $[-\pi, \pi]$, so that all four quadrants are used. The average angle can be calculated with

$$\langle \theta \rangle = \text{ATAN2}(\langle p_y \rangle, \langle p_x \rangle). \tag{2.24}$$

This procedure avoids the pitfall associated with Equation 2.18.

In Chapter 7, we discuss Euler's identity for the complex exponential

$$e^{i\theta} = \cos \theta + i \sin \theta, \tag{2.25}$$

where $i = \sqrt{-1}$. Many fields of study, including electrical engineering and optics, make wide use of the complex exponential. Equation 2.25 provides a particularly convenient mathematical framework to perform the calculation of Equations 2.22 and 2.23. Specifically, we calculate the average complex exponential

$$\langle z \rangle = \frac{1}{N} \sum_{k=1}^{N} e^{i\theta_k} \tag{2.26}$$

and then using Equation 2.25 associate the average phasor component with its real and imaginary parts:

$$\begin{aligned} \langle p_x \rangle &= \text{Re}\langle z \rangle, \\ \langle p_y \rangle &= \text{Im}\langle z \rangle. \end{aligned} \tag{2.27}$$

As illustrated in exercise 2.5, the vector and complex exponential methods are closely related to each other because they both work with the trigonometric functions sine and cosine, instead of dealing with angles directly. Both methods are equally valid to calculate the average of a set of angles.

For some applications, it is useful to generate a visual plot of angular values, known as a *phase map*. At each point (x, y) in the phase map, the value of the phase angle ϕ is depicted as a grayscale value, i.e., the brightness of the image is proportional to the value of $\phi(x, y)$. Often, such phase maps are generated from measured data using an inverse trigonometric function, such as ATAN2.

The $2\pi n$ ambiguity in the value of an angle can complicate the interpretation of a phase map. The limited output range of inverse trigonometric functions can cause discontinuities in the phase map whenever $|\phi|$ exceeds a certain threshold. These discontinuities are known as *phase wraps*. For example, if the phase map is generated using the ATAN2 function, phase wraps occur when $|\phi| > \pi$ because the output of the ATAN2 function is limited to the range $[-\pi, \pi]$.

Figure 2.3a shows a phase map image acquired from a magnetic resonance imaging test measurement of an object that approximately simulates the shape of a human head and neck. The phase map in Fig. 2.3a, which was generated using the ATAN2 function, displays phase wraps that appear as abrupt transitions between white and black regions of the image, corresponding to the boundary between positive and negative values of ϕ. Because a total of six phase wraps are apparent from left to right in the phase map in Fig. 2.3a, the actual range of values of ϕ covered in the phase map (i.e., the true dynamic range) is approximately $6 \times 2\pi = 12\pi$.

The undesirable phase wraps can be removed by *phase unwrapping*, for which many algorithms have been developed (Ghiglia and Pritt 1998; a detailed discussion of phase unwrapping algorithms is provided in this book). If $-\pi \leq \phi \leq +\pi$ in the wrapped phase map, these algorithms work by adding or subtracting an appropriate multiple of 2π at each location in the phase map so as to make ϕ a continuous function of x and y. Figure 2.3b shows the corresponding phase map where the phase in the test object has been unwrapped. Observe how Fig. 2.3b provides a truer depiction of the spatial dependence of $\phi(x, y)$ compared with Fig. 2.3a. Finally, we note that the construction of Equation 2.15 is an example of one-dimensional phase unwrapping. In that particular case, however, we unwrapped the phase by adding π rather than a multiple of 2π, because we were dealing with the two-quadrant arctangent function instead of ATAN2.

The ambiguity of the value of an angle also has deeper implications in fields such as quantum mechanics, where certain dynamical variables such as position and momentum correspond to measurable quantities called *observables*. Because an angle θ can only be defined modulo 2π radians, it is *not* an observable. As described in Peierls (1979; further description of the pitfalls associated with defining an angle operator in quantum mechanics is provided in this book), constructing a mathematical operator in quantum mechanics corresponding to the angle θ can lead to contradictory results. Instead, the problem can be avoided by performing the intermediate steps of a quantum mechanical calculation with the exponential operator $e^{i\theta}$, analogous to the use of complex exponentials in Equation 2.26.

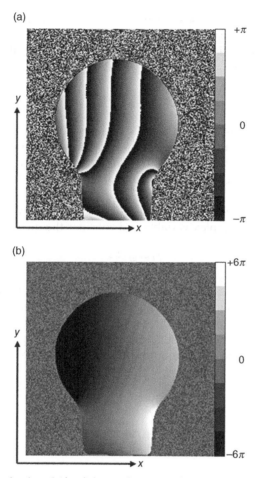

FIG. 2.3. A grayscale plot of $\phi(x, y)$, i.e., a phase map, from a magnetic resonance image of a test object that is shaped like a human head and neck. (a) The value of phase is limited to the range $[-\pi, \pi]$, which causes abrupt transitions or phase wraps. (b) An unwrapped phase map depicts the true continuous nature of the underlying function $\phi(x, y)$ obtained from the test object. Note that the range of values in (b) is extended to $[-6\pi, +6\pi]$.

There are still other situations where the ambiguity in the value of variable can cause difficulties. Consider the relationship

$$x^2 = 4, \tag{2.28}$$

which is satisfied when $x = -2$ or $x = +2$. Consequently, the inverse of the squaring operation, $y = x^{1/2}$, can mean either $y = +\sqrt{x}$ or $y = -\sqrt{x}$. This introduces ambiguity in much the same way as the multiple branches of the inverse tangent discussed earlier.

This ambiguity often arises when working with a function of a complex variable. The problem can be aggravated by a common use of the word "function" in complex analysis. We typically do not refer to the relationship between real variables x and y,

$$y = x^{1/2}, x > 0, \tag{2.29}$$

as a "function" because two values of y correspond to each value of x, i.e., y is multivalued. In complex analysis, however, the terminology is not as standardized, and we often encounter the term "multivalued function." This can cause confusion, so dealing with multivalued complex functions requires special care.

The following discussion might be useful to those who have studied complex analysis. Let z be a complex variable and consider the quantity

$$f(z) = z^{1/2}. \tag{2.30}$$

As outlined in exercise 2.8, we can express z and $f(z)$ in terms of the polar coordinates ρ and θ. Suppose we start at the point z_0 and make one complete circuit in the complex plane by increasing the value of θ by 2π, as illustrated in Fig. 2.4. Although the new value of z is the same complex number as the starting point, i.e., it has the same real and imaginary parts as z_0, the function

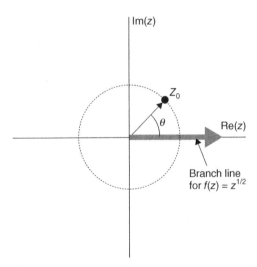

FIG. 2.4. A point z_0 lies in the complex plane and is associated with the polar angle θ. As described in the text and in exercise 2.8, certain functions, such as $f(z) = z^{1/2}$ can return ambiguous results depending on how many integer multiples of 2π are added to the angle θ. To avoid this pitfall, a boundary called a branch line (thick, gray arrow) can be constructed in the complex plane.

$f(z)$ changes sign. As also outlined in exercise 2.8, the original sign of $f(z_0)$ is restored after a second complete circuit. These sign changes can cause confusion and errors when working with $f(z)$.

Several standard techniques have been developed to avoid this pitfall. One method is to simply restrict the variation of z by constructing a barrier in the complex plane called a branch cut or *branch line*. In the case of Equation 2.30, we can construct our branch line from 0 to ∞ along the real axis, as indicated in Fig. 2.4. The end point $z = 0$ is called a branch point. Provided our path in the complex plane does not cross this branch line, the function $f(z) = z^{1/2}$ remains single valued.

An alternative, but related approach, removes the branch-line restriction and instead defines a *Riemann surface* that contains multiple sheets. These surfaces are named after the nineteenth-century German mathematician Bernhard Riemann. On any particular sheet of the Riemann surface, the complex function is single valued. The complex function $f(z) = z^{1/2}$ contains two sheets, while $f(z) = z^{1/n}$ contains n sheets. Consider the case of $f(z) = z^{1/2}$ and the multiple circuits in the complex plane previously described. During the first circuit in the complex plane, when θ reaches 2π (i.e., where the branch line would have been encountered), we can continue onto the second sheet corresponding to values of θ running from 2π to 4π. Continuing with the second circuit, when θ exceeds 4π, we return to the first sheet. The geometric construct of the Riemann surface can be quite convenient because it allows us to define a multivalued function in a continuous manner. Riemann surfaces are described in greater detail in many texts on complex analysis, such as Gamelin (2003).

The next pitfalls we discuss are related to the topic of *notation*. Notation is a general term covering the choice of symbols to represent variables and operators that appear in an equation. It also refers to the related format of the symbols such as the use of subscripts, primes, arrows representing vectors, and choice of font.

Poor choice of notation can cause confusion and errors in equations. Sometimes, the problem is purely superficial, i.e., the equation is correct but a poor choice of notation makes its interpretation more difficult. In other instances, notational problems can lead to substantive errors, and incorrect numerical results.

As a common example of the more substantive type of error in notation, consider the misuse of a dummy variable such as an integration variable or a summation index. The integration variable t in the equation

$$g(x) - g(a) = \int_a^x f(t)\,dt \tag{2.31}$$

can alternatively be chosen to be x', v or virtually any other symbol, *except* the variable x. The symbol x is reserved for the upper limit of integration. This choice ensures that the derivative of the integral with respect to x is

$$\frac{dg}{dx} = f(x). \tag{2.32}$$

This idea can be generalized to the situation where the upper integration limit is itself any differentiable function $u(x)$ by using the chain rule for differentiation:

$$\frac{d}{dx}\left(\int_a^{u(x)} f(t)\,dt\right) = f(u(x))\cdot\frac{du}{dx}. \tag{2.33}$$

If we want to calculate the square of the function g by replicating the right-hand side of Equation 2.31, it is important *not* to reuse the same dummy integration variable. That is, we can correctly express the quantity $[g(x) - g(0)]^2$ in terms of the double integral

$$[g(x)-g(0)]^2 = \int_0^x f(t)\,dt \times \int_0^x f(w)\,dw = \int_0^x\int_0^x f(t)\,f(w)\,dw\,dt. \tag{2.34}$$

But reusing the same integration variable t yields the ambiguous expression

$$[g(x)-g(0)]^2 = \int_0^x f(t)\,dt \times \int_0^x f(t)\,dt \quad [\text{incorrect}]. \tag{2.35}$$

If we try to write the right-hand side of Equation 2.35 as a double integral as in Equation 2.34, the resulting expression is meaningless. It is not clear with which integral the symbol t is associated. The importance of not reusing the same symbol for integration variables is also illustrated by exercise 2.13.

We also have to be careful about the choice of summation indices, such as those used to express matrix multiplication. If we are given an $N \times N$ matrix $\mathbf{M}$, we write its ij-th element as M_{ij}. If we want to find the ij-th element of the matrix product $\mathbf{M}^2$, we use the rules of matrix multiplication:

$$\left(\mathbf{M}^2\right)_{ij} = \sum_{k=1}^N M_{ik} M_{kj}. \tag{2.36}$$

We have chosen the dummy summation index to be k, but any symbol *except i or j* is acceptable. (Note that as k takes on the different values indicated by the summation, it can assume the same numerical value as i or j.) As illustrated in exercise 2.4, using an improper summation index can lead to incorrect numerical results.

The choice of notation is an important aspect of any equation. Even though a superficial or "cosmetic" error might not directly cause an incorrect numerical result, it can seriously hinder our understanding of an equation. We are accustomed to using the symbol x to represent an independent variable, and we often refer the abscissa of a graph as the "x-axis." Letters such as a, b, and c often represent constants. As an example of a superficial problem with notation, consider the familiar quadratic equation and its solution. Assuming that $a \neq 0$, we have

$$ax^2 + bx + c = 0, \qquad (2.37)$$

with the solution

$$x = \frac{-b \pm \sqrt{b^2 - 4ac}}{2a}. \qquad (2.38)$$

Now consider an equivalent pair of equations constructed using a nonstandard choice for the notation where a represents the independent variable and x, c, and b represent the constants:

$$xa^2 + ca + b = 0. \qquad (2.39)$$

Equation 2.39 has the solution

$$a = \frac{-c \pm \sqrt{c^2 - 4bx}}{2x}. \qquad (2.40)$$

Equation 2.40 contains the same information as Equation 2.38 in the sense that a computer will execute instructions describing the two equations equivalently. For human readers, however, Equation 2.38 is familiar, while Equation 2.40 seems discordant and distracting. We might find ourselves mentally translating Equation 2.40 back to the familiar form of Equation 2.38.

The case illustrated by the quadratic formula is extreme, and there certainly is flexibility when choosing good notation. Often, the trade-off to be weighed is between clarity and compactness. For example, in the study of general relativity, the resulting expressions contain numerous matrix and tensor products, so a simplifying notational convention is adopted. To avoid writing many "Σ" signs, summation is implied whenever a repeated index is encountered, for example, the index k in Equation 2.36. Workers in this field become accustomed to this alternative convention and could write that equation compactly as "$(\mathbf{M}^2)_{ij} = M_{ik}M_{kj}$." When this convention is adopted, a special note is made for any cases where an index is repeated but is not to be summed over.

In most cases, it is good practice to adopt the most standard choices for notation that are available. The use of standard notation in equations usually does not stifle creativity but instead can have the opposite effect because its use frees us from the burden of making mental translations as in Equation 2.40.

2.2   A FEW USEFUL TRICKS

There are several useful tricks for evaluating the mean of a set of variables and related quantities like the variance. Suppose that a variable U takes on one of the discrete values $U_1, U_2, U_3, \ldots U_N$. Let the probability that U takes the value U_i be denoted by $p(i)$. Because the total probability is normalized to the value of 1,

$$\sum_{i=1}^{N} p(i) = 1, \tag{2.41}$$

then the mean or expected value of U, denoted $\langle U \rangle$, is defined by the sum

$$\langle U \rangle = \sum_{i=1}^{N} p(i) U_i. \tag{2.42}$$

We will consider an interesting and important special case where the probability distribution is proportional to the exponential function

$$p(i) \propto e^{-\beta U_i}, \tag{2.43}$$

where β is a parameter that does not depend on the index i. With the normalization condition of Equation 2.41, the probability distribution becomes

$$p(i) = \frac{e^{-\beta U_i}}{\sum_{j=1}^{N} e^{-\beta U_j}}. \tag{2.44}$$

EXAMPLE 2.2

Construct an expression for (a) the mean and (b) the variance of the variable U when its probability distribution is given in Equation 2.44. Express the results in terms of derivatives and a single summation.

ANSWER

(a) We begin by defining the function Q that is used to normalize the probability

$$Q(\beta) \equiv \sum_{i=1}^{N} e^{-\beta U_i}. \tag{2.45}$$

Combining Equations 2.42, 2.44, and 2.45 yields

$$\langle U \rangle = \frac{1}{Q} \sum_{i=1}^{N} U_i e^{-\beta U_i}. \tag{2.46}$$

The first trick is to recognize that the numerator of the right-hand side of Equation 2.46 can be written in terms of a partial derivative of Q with respect to β:

$$\frac{1}{Q}\sum_{i=1}^{N}U_i e^{-\beta U_i} = \frac{1}{Q}\left(-\frac{\partial}{\partial\beta}\left(\sum_{i=1}^{N}e^{-\beta U_i}\right)\right) = -\frac{1}{Q}\frac{\partial Q}{\partial\beta}. \qquad (2.47)$$

Note that the derivative with respect to β can be brought inside the summation because β does not depend on the index i.

The second trick is to recall that, according to the chain rule, the derivative of the natural logarithm of a function is given by

$$\frac{d}{dx}(\ln[f(x)]) = \frac{f'(x)}{f(x)}, \qquad (2.48)$$

provided that $f > 0$. If f is a function of multiple variables, then we can generalize Equation 2.48 with the use of a partial derivative, i.e.,

$$\frac{\partial}{\partial x}(\ln[f(x, y)]) = \frac{\partial f(x, y)}{\partial x} \times \frac{1}{f(x, y)}. \qquad (2.49)$$

Comparing Equations 2.47 and 2.49, we can conclude that the expected mean value of U can be written compactly as

$$\langle U \rangle = -\frac{\partial(\ln Q)}{\partial\beta}. \qquad (2.50)$$

As was described in Chapter 1, a similar technique can be used to evaluate some other definite and indefinite integrals.

(b) The variance $(\Delta U)^2$ is defined by the following sum:

$$(\Delta U)^2 = \sum_{i=1}^{N}(U_i - \langle U \rangle)^2 p(i). \qquad (2.51)$$

The variance can be found using the equivalent expression (exercise 2.5)

$$(\Delta U)^2 = \langle U^2 \rangle - \langle U \rangle^2. \qquad (2.52)$$

The expected value of U^2 is given by

$$\langle U^2 \rangle = \frac{1}{Q}\sum_{i=1}^{N}U_i^2 e^{-\beta U_i} = \frac{1}{Q}\left(\frac{\partial^2}{\partial\beta^2}\sum_{i=1}^{N}e^{-\beta U_i}\right) = \frac{1}{Q}\frac{\partial^2 Q}{\partial\beta^2}. \qquad (2.53)$$

Evaluating Equation 2.52 using Equations 2.50 and 2.53, we find that the variance is

$$(\Delta U)^2 = \frac{1}{Q}\frac{\partial^2 Q}{\partial\beta^2} - \frac{1}{Q^2}\left(\frac{\partial Q}{\partial\beta}\right)^2. \qquad (2.54)$$

It is straightforward to show that Equation 2.54 can be expressed even more compactly as the second derivative of the natural logarithm of Q:

$$(\Delta U)^2 = \frac{\partial^2 \ln(Q(\beta))}{\partial \beta^2}. \tag{2.55}$$

The relationships derived in Example 2.2 have useful applications in the field of statistical mechanics. There, the parameter β is inversely proportional to the absolute temperature T (measured in kelvins) of the system:

$$\beta = \frac{1}{kT}, \tag{2.56}$$

where k is Boltzmann's constant, which is approximately 1.38065×10^{-23} J/K. The constant k is named in honor of the Austrian physicist Ludwig Boltzmann, who worked in the nineteenth and early twentieth centuries. The quantities U_i labeled by the index i represent the possible energy values of the system. Consequently, the ratio $U_i/kT = \beta U_i$ is a dimensionless quantity. In the statistical mechanical context, the function $Q(\beta)$ defined in Equation 2.45 is called the *partition function*. The exponential probability distribution given in Equation 2.43 is known as Boltzmann weighting.

Because both the mean and variance of the variable U can be obtained by differentiating the same partition function Q, another useful relationship can be established. From Equation 2.56 and the chain rule for differentiation, it follows that

$$\frac{\partial f}{\partial \beta} = -kT^2 \frac{\partial f}{\partial T}. \tag{2.57}$$

Combining Equations 2.50, 2.55, and 2.57, we obtain

$$(\Delta U)^2 = kT^2 \frac{d\langle U \rangle}{dT}. \tag{2.58}$$

The variance $(\Delta U)^2$ is a measure of the fluctuation of the energy for the system being described by statistical mechanics. That is, when $(\Delta U)^2 = 0$, the value of the energy remains constant at its mean $\langle U \rangle$. This condition is met as $\beta \to \infty$, or equivalently as the absolute temperature T approaches zero. On the other hand, $(\Delta U)^2 > 0$ suggests that the system takes on a variety of values of energy, characterized by the range ΔU centered about the mean energy.

Equation 2.58 provides a useful link between the fluctuation in the energy and rate of change of the average energy. It is an example of a *fluctuation–dissipation* relationship. We will see other versions of this type of relationship when we discuss equations related to Brownian motion and related phenomena in Chapter 5.

The next useful trick deals with the parametric representation of equations. Frequently, we are presented with coupled equations, and a standard technique is to combine them in order to eliminate one or more of the variables. An example is the procedure of forming a linear combination of a set of equations. As a simple illustration, we can add

$$x + y = 3,$$
$$x - y = 7, \tag{2.59}$$

to eliminate the variable y. We then obtain $2x = 10$, from which it follows that $x = 5$ and $y = -2$.

The opposite strategy of separating a single equation into multiple parts can be quite useful too. This is accomplished by introducing an additional parameter, and the resulting equations are called a *parametric* representation. For example, this representation is useful for describing motion of a projectile, such as a thrown ball. Neglecting air resistance, the ball follows a parabolic trajectory in the xy-plane, where the y-axis is along the vertical direction, and x is a horizontal direction:

$$y = ax^2 + bx + c. \tag{2.60}$$

Equation 2.60 describes the path of the ball, but it does not tell where it will be at a specified time. Therefore, it is usually more convenient to use the parametric representation

$$x(t) = x_0 + v_{0x}t,$$
$$y(t) = y_0 + v_{0y}t - \frac{1}{2}gt^2, \tag{2.61}$$

where the parameter t represents time, g is the acceleration due to gravity, and the constants with the subscript "0" represent the projectile's initial position or velocity. The constants a, b, and c appearing in Equation 2.60 can be expressed in terms of the constants appearing in Equation 2.61 by eliminating the parameter t (see exercise 2.7).

The following example illustrates the conversion of a single, second-order differential equation into two linked, first-order differential equations by introducing a parametric function.

EXAMPLE 2.3

Consider the second-order differential equation

$$\frac{d^2y}{dx^2} = r\sqrt{1 + \left(\frac{dy}{dx}\right)^2}, \tag{2.62}$$

where r is a nonzero constant. Solving Equation 2.62 for y appears difficult because its second derivative is expressed as a nonlinear function of its first derivative.

(a) Parameterize Equation 2.62 so that two first-order, coupled differential equations are obtained.
(b) Solve the two coupled differential equations subject to the boundary conditions $y(0) = 0$ and $y'(0) = 0$.

ANSWER

(a) We introduce a new parametric function $q(x)$

$$\frac{dy}{dx} = q(x). \tag{2.63}$$

Equation 2.63 is one of the parametric equations and its pair is

$$\frac{dq}{dx} = r\sqrt{1+q^2}. \tag{2.64}$$

Simple inspection confirms that Equations 2.63 and 2.64 are equivalent to the original second-order equation, Equation 2.62.

(b) By solving the pair of first-order equations with boundary conditions, we obtain the desired function $y(x)$ satisfied by Equation 2.62. The boundary condition $y'(0) = 0$ is equivalent to $q(0) = 0$, which provides the required boundary conditions for the differential equation given by Equation 2.64.

Equation 2.64 is a separable differential equation, so to solve for q, we evaluate the integral

$$\int_{q(0)}^{q} \frac{du}{\sqrt{1+u^2}} = rx. \tag{2.65}$$

Using a table of integrals and applying the boundary condition $q(0) = 0$, Equation 2.65 yields the inverse hyperbolic sine

$$\sinh^{-1} q = rx. \tag{2.66}$$

Substituting $q = \sinh rx$ into Equation 2.63 gives

$$\frac{dy}{dx} = \sinh(rx). \tag{2.67}$$

Equation 2.67 is then integrated to yield a hyperbolic cosine

$$y(x) = \frac{1}{r}\cosh(rx) + C. \tag{2.68}$$

Finally, from the boundary condition $y(0) = 0$, we determine that $C = -1/r$, so the solution to Equation 2.62 with the specified boundary conditions is

$$y(x) = \frac{1}{r}(\cosh(rx) - 1). \tag{2.69}$$

The function in Equation 2.69 arises when we study the problem of a chain or rope that hangs freely while supported at both ends. As explored in more detail in Chapter 8, Example 8.2, the resulting shape is known as a catenary.

2.3 A FEW "ADVANCED" TRICKS*

The Dirac delta function or *delta function* $\delta(x)$ is zero everywhere *except* at the single point $x = 0$. It can be defined by the relationship

$$\delta(x) = \begin{cases} 0, x \neq 0 \\ \int\limits_{-\infty}^{\infty} \delta(x)\,dx = 1 \end{cases}. \tag{2.70}$$

To satisfy Equation 2.70, the amplitude of $\delta(x)$ must be infinitely large at the origin. Consequently, it is not a standard function and is sometimes called the Dirac delta "distribution." The Dirac delta function $\delta(x)$ is named in honor of the twentieth-century English physicist Paul Adrien Maurice Dirac.

The use of the delta function greatly simplifies the evaluation of many integrals. The delta function has the *sifting* property that it selects out a single point from an integrand as follows

$$\int\limits_{-\infty}^{\infty} \delta(x - a)f(x)\,dx = f(a). \tag{2.71}$$

which is consistent with the fact that $\delta(x-a)$ is zero everywhere except at $x = a$. Another useful property is

$$\int\limits_{-\infty}^{x} \delta(t)\,dt = H(x) \text{ or } \delta(x) = \frac{dH}{dx}, \tag{2.72}$$

*This section (and the related exercises 2.9–2.13) use somewhat more advanced mathematical concepts. This material can be skipped by readers who are not currently interested in Dirac delta functions and Fourier transforms.

where H is the unit step function:

$$H(x) = \begin{cases} 0, & x < 0 \\ 1, & x > 0 \end{cases}.$$ (2.73)

In the two most commonly adopted conventions, the value of $H(0)$ is defined to be 1 or 0.5. From Equation 2.73, the unit step function is discontinuous at the origin, so Equation 2.72 does not apply for the point $x = 0$. The unit step function is also called the Heaviside function, after the English engineer, mathematician, and physicist Oliver Heaviside. Equation 2.72 can be generalized to

$$\int_{-\infty}^{x} \delta(t-a) f(t) dt = f(a) H(x-a),$$ (2.74)

provided that $x \neq a$. Equations 2.71 and 2.74 provide a few examples of how the Dirac delta function can simplify the evaluation of integrals. Delta and step functions prove useful in many areas of studies. For example, the step function $H(t - a)$ is a convenient way to describe the switching of an electrical circuit to the "on" state at the time $t = a$. Similarly, point sources that are observed in nature such as the electrical charge q on an electron are well modeled by delta functions. For example, we can describe the charge density ρ (charge per unit length) of an idealized point charge q located at the origin by the expression

$$\rho(x) = q\delta(0).$$ (2.75)

With the use of relationships such as Equation 2.72, quantities such as $\rho(x)$ can be manipulated to solve certain differential equations (see exercise 2.10). It is also straightforward to generalize Equation 2.75 to the three-dimensional case, i.e., charge per unit volume.

Delta functions can be expressed in a variety of alternative forms, which further increases their utility. Equation 2.70 suggests that a representation for $\delta(x)$ is found by constructing an expression that is infinitely peaked at $x = 0$, while being normalized to maintain unit area. The simplest such representation is a "box" that approaches infinite height and zero width:

$$\delta(x) = \lim_{a \to 0+} \begin{cases} \dfrac{1}{a}, & |x| \leq a/2 \\ 0, & |x| > a/2 \end{cases}.$$ (2.76)

Other representations of the Dirac delta function can be constructed using similar reasoning, and three widely used examples are

$$\delta(x) = \lim_{\sigma \to 0+} \frac{1}{\sigma\sqrt{2\pi}} e^{\frac{-x^2}{2\sigma^2}},$$ (2.77)

$$\delta(x) = \lim_{\lambda \to 0+} \frac{2\lambda}{\lambda^2 + (2\pi x)^2},$$

(2.78)

and

$$\delta(x) = \lim_{a \to \infty} \frac{\sin 2\pi x a}{\pi x}.$$

(2.79)

(See exercise 2.9.) Perhaps, the most useful representation, however, is the definite integral

$$\delta(a) = \frac{1}{2\pi} \int_{-\infty}^{\infty} e^{\pm iau} du.$$

(2.80)

A derivation of Equation 2.80 is outlined in exercise 2.11. The discrete analog of Equation 2.80 is described in exercise 2.12.

Equation 2.80 has many useful applications. It provides a straightforward way to establish the Fourier integral theorem (see exercise 2.13). If $G(t)$ is the Fourier transform of $g(\omega)$,

$$G(t) = \int_{-\infty}^{\infty} g(\omega) e^{-i\omega t} d\omega,$$

(2.81)

then the Fourier integral theorem states that we can recover the original function by performing the inverse Fourier transform:

$$g(\omega) = \frac{1}{2\pi} \int_{-\infty}^{\infty} G(t) e^{+i\omega t} dt.$$

(2.82)

The next example illustrates another use of the integral representation of the delta function. In some equations, the integration variables are linked together by a constraint, which can make the integral difficult to evaluate. In the following example, we describe an approach based on Equation 2.80 that can alleviate this difficulty.

EXAMPLE 2.4

Consider a one-dimensional random walk composed of a sequence of N statistically independent steps. The probability of moving a distance between s and $s + ds$ during each step is given by $\rho(s)ds$, where $\rho(s)$ is a probability density function.

What is the net probability density $P_N(x)$ that after N steps, the net result of the random walk brings us to the vicinity of x?

ANSWER

This problem involves the accumulation of N individual steps s_i subject to a constraint on their sum:

$$\sum_{i=1}^{N} s_i = x. \tag{2.83}$$

If we consider a single step s_1, we obtain the simple result

$$P_1(x) = \rho(x) \tag{2.84}$$

subject to the constraint that $x = s_1$.

For a random walk composed of two steps, the joint probability distribution is obtained by simple multiplication of the two individual distributions because the two steps are assumed to be statistically independent. The constraint that $x = s_1 + s_2$ is expressed as $s_2 = x - s_1$ and incorporated directly into the argument of one of the probability density functions, yielding

$$P_2(x) = \int_{s_1=-\infty}^{\infty} \rho(s_1)\rho(x-s_1)\,ds_1 \tag{2.85}$$

subject to the constraint that $x = s_1 + s_2$. The question now arises: how to generalize Equations 2.84 and 2.85 to an arbitrary number N steps? As discussed in more detail in Chapter 7, an effective method to generalize equations is to look for a pattern. Based on the apparent pattern, we can generalize Equations 2.84 and 2.85 to

$$P_N(x) = \int_{s_1=-\infty}^{\infty} \int_{s_2=-\infty}^{\infty} \ldots \int_{s_{N-1}=-\infty}^{\infty} \rho(s_1)\rho(s_2)\ldots\rho\left(x - \sum_{i=1}^{N-1} s_i\right) ds_1 ds_2 \ldots ds_{N-1}, \tag{2.86}$$

which is consistent with the constraint of Equation 2.83.

It might not be clear how to evaluate the complicated multiple integral expressed in Equation 2.86. An interesting path to the solution is to realize that the constraint of Equation 2.83 can be enforced with a delta function. Using the sifting property of the delta function in Equation 2.71, Equation 2.84 can be recast into the form

$$P_1(x) = \int_{s_1=-\infty}^{\infty} \rho(s_1)\delta(x-s_1)\,ds_1. \tag{2.87}$$

Similarly, Equation 2.85 can be recast as a double integral with a delta function enforcing the constraint

$$P_2(x) = \int_{s_1=-\infty}^{\infty} \int_{s_2=-\infty}^{\infty} \rho(s_1)\rho(s_2)\delta(x-(s_1+s_2))\,ds_1 ds_2. \tag{2.88}$$

The pattern emerging from Equations 2.87 and 2.88 is becoming clearer. The desired generalization is

$$P_N(x) = \int\limits_{s_1=-\infty}^{\infty} \int\limits_{s_2=-\infty}^{\infty} \cdots \int\limits_{s_{N-1}=-\infty}^{\infty} \rho(s_1)\rho(s_2)\ldots\rho(s_N)\delta\left(x - \sum_{i=1}^{N} s_i\right) ds_1 ds_2 \ldots ds_N.$$

(2.89)

From the integral representation of the delta function, Equation 2.80, we can express

$$\delta\left(x - \sum_{i=1}^{N} s_i\right) = \frac{1}{2\pi} \int\limits_{k=-\infty}^{+\infty} e^{-ik\left(x - \sum_{i=1}^{N} s_i\right)} dk.$$

(2.90)

Substituting Equation 2.90 into Equation 2.89, we obtain

$$P_N(x) = \frac{1}{2\pi} \int\limits_{k=-\infty}^{+\infty} \int\limits_{s_1=-\infty}^{\infty} \int\limits_{s_2=-\infty}^{\infty} \cdots \int\limits_{s_N=-\infty}^{\infty} e^{-ikx}\left[\rho(s_1)e^{is_1k}\right]\left[\rho(s_2)e^{is_2k}\right]$$

$$\ldots\left[\rho(s_N)e^{is_Nk}\right] ds_1 ds_2 \ldots ds_N dk.$$

(2.91)

At first glance, Equation 2.91 still looks very complicated. It can be simplified greatly, however, because each integral over a variable s_i can be evaluated separately, and each of those integrals yields the identical result. We define

$$\phi(k) \equiv \int\limits_{s=-\infty}^{+\infty} e^{iks}\rho(s)\,ds,$$

(2.92)

which from Equation 2.82 is proportional to the inverse Fourier transform of the probability density $\rho(s)$. From Equations 2.91 and 2.92, we obtain the desired result

$$P_N(x) = \frac{1}{2\pi} \int\limits_{k=-\infty}^{+\infty} e^{-ikx}\left[\phi(k)\right]^N dk.$$

(2.93)

Equations 2.92 and 2.93 together are considerably more manageable than Equation 2.89, so they provide a welcome simplification. Equation 2.93 will not be too surprising to readers who are familiar with the convolution operation. They will recognize Equation 2.87 as the convolution of two probability density distributions. The convolution of two functions in one domain (i.e., the s-domain) is equivalent to multiplication of their two Fourier transforms in the conjugate domain (i.e., the k-domain). That result leads to the factor $[\phi(k)]^N$ that appears in Equation 2.93.

One of the most remarkable and important results in the study of probability is the central limit theorem. Applied to the random walk described in

Example 2.4, it states that as the number of steps N grows large, the net probability distribution $P_N(x)$ tends toward a normal or Gaussian distribution, independent of the specific shape of $\rho(s)$. The distribution is named after the well-known German mathematician and physicist Carl Friedrich Gauss. Specifically, as N increases,

$$P_N(x) \to \frac{1}{\sigma\sqrt{2\pi}} e^{-\frac{(x-\langle x \rangle)^2}{2\sigma^2}}, \tag{2.94}$$

where the mean $\langle x \rangle$ and standard deviation σ for the entire random walk can be expressed in terms of the mean and standard deviation of the individual steps by

$$\langle x \rangle = N\langle s \rangle = N \int_{-\infty}^{\infty} \rho(s)s\,ds \tag{2.95}$$

and

$$\sigma^2 = N(\Delta s)^2 = N\left[\langle s^2 \rangle - \langle s \rangle^2\right] = N\left[\int_{-\infty}^{\infty} \rho(s)s^2\,ds - \langle s \rangle^2\right]. \tag{2.96}$$

A derivation of Equations 2.94–2.96 is provided in standard texts such as Reif (1965; a more detailed description of the connection between the random walk problem outlined in Example 2.4 and the central limit theorem is provided in this book). We will only make a few general comments here. A key feature of the derivation is that the function $\phi(k)$ is dominated by small values of k, due to the oscillating factor e^{iks} in the integrand of Equation 2.92. Those oscillations become increasingly rapid as k increases. Also, for large values of N, the maximum of the function $\phi(k)$ in the integrand of Equation 2.93 becomes increasingly accentuated through repeated multiplication. These two facts justify the use of the power series expansion in the variable k:

$$[\phi(k)]^N = e^{N\ln\phi(k)} \approx \exp\left(N\left(i\langle s \rangle k - \frac{1}{2}\Delta s^2 k^2\right)\right). \tag{2.97}$$

The Gaussian form of Equation 2.94 results when the approximation of Equation 2.97 is substituted into Equation 2.93. Further discussion about related methods to approximate definite integrals is presented in Chapter 5.

EXERCISES

(2.1) When a series of discrete points is arranged on a line, we either can count the points or else count the intervals between the points. A very basic, yet commonly committed, error is to confuse these two quantities.

(a) How many birthdays do you have from the day of your tenth birthday to the day of your twentieth birthday, including those two? How many years elapse? (Assume your birthday does not fall on February 29.)

(b) Show that the fractional error E associated with confusing n points and the intervening intervals is

$$E = \frac{|n-(n-1)|}{n} = \frac{1}{n}.$$

(c) In science and engineering, we are trained to carefully check our work. Sometimes, errors of intermediate scale are the most troublesome. For each of the following values: $n = 5$, $n = 500$, and $n = 5 \times 10^{10}$, comment on how significant the relative error E is and also how likely it is that such an error might escape detection.

(2.2) Show that

$$G(x) = \begin{cases} \dfrac{1}{2}x^2, \ x < \dfrac{1}{2} \\ \dfrac{1}{2}x^2 + 1, \ x \geq \dfrac{1}{2} \end{cases}$$

is an antiderivative of $g(x) = x$ everywhere *except* at $x = 1/2$. Can we use $G(x)$ to calculate $\int_0^1 g(x)dx$? Explain.

(2.3) Apply Equation 2.24 to calculate the average of $\theta_1 = 170°$ and $\theta_2 = -170°$.

(2.4) Consider the 2×2 matrix $\mathbf{M} = \begin{pmatrix} 1 & 5 \\ 3 & 2 \end{pmatrix}$. Calculate the following:

(a) the ij-th element of the matrix $\mathbf{MM} = \mathbf{M}^2$;

(b) the squares of the ij-th element of $\mathbf{M}$, or $(M_{ij})^2$.

(c) Comment on the role of a dummy summation index in part (a) to ensure that the results in parts (a) and (b) differ.

(2.5) Show that

$$(\Delta U)^2 = \langle (U - \langle U \rangle)^2 \rangle = \langle U^2 - 2U \langle U \rangle + \langle U \rangle^2 \rangle$$
$$= \langle U^2 \rangle - \langle U \rangle^2.$$

(2.6) Suppose that a two-dimensional vector $\vec{v} = (v_x, v_y)$ is expressed in polar coordinates as

$$v_x = v_0 \cos\theta,$$
$$v_y = v_0 \sin\theta.$$

(a) Show that $\vec{v}$ is a unit vector, i.e., has a magnitude equal to 1, if and only if $v_0 = 1$. Sometimes unit vectors are denoted with a caret, such as $\hat{v}$.

(b) Show that

$$e^{i\theta} = \frac{v_x + iv_y}{\sqrt{v_x^2 + v_y^2}}.$$

(c) Comment on the differences and similarities between the complex exponential $e^{i\theta}$ and the unit vector $\hat{v} = (\cos\theta, \sin\theta)$.

(2.7) Consider the parametric representation of the projectile motion given in Equation 2.61. Eliminate the parameter t, and solve for the constants a, b, and c that appear in Equation 2.60 in terms of the constants that appear in Equation 2.61.

(2.8) Let $f(z) = z^{1/2}$, where z is a complex variable.

(a) Using the polar form $z = \rho e^{i\theta}$, show that $f(z) = \sqrt{\rho}e^{i\theta/2}$.

(b) Suppose we start at the point $z_0 = \rho_0 e^{i\theta_0}$ (with $\rho_0 \neq 0$) and make one complete circular, counterclockwise revolution in the complex plane about the origin and return to z_0. Using the result in part (a), show that the value of function is negated, that is,

$$f(z_0) \rightarrow \rho_0 e^{i(\theta_0 + 2\pi)/2} = -f(z_0).$$

(c) Show that if we then make a second, counterclockwise revolution in the complex plane and again return to z_0, the original sign of $f(z_0)$ is restored.

(2.9) Graph Equation 2.77 for $\sigma = 1, 0.1$, and 0.01. Graph Equation 2.78 for $\lambda = 1, 0.1$, and 0.01. Comment on the similarities between these sets of graphs and their relationship to the Dirac delta function.

(2.10) Consider the differential equation

$$\frac{d^2 y}{dx^2} = q\delta(0),$$

with the conditions that q is a constant, $y(x) = 0$ for $x < 0$, $y(x)$ is continuous for all values of x, and $y'(x)$ exists for all x *except* $x = 0$. Find $y(x)$. Hint: argue that the first derivative $y'(x)$ is proportional to the unit step function.

(2.11)

(a) Verify that $\displaystyle\int_{-\infty}^{\infty} \frac{\sin u}{u}\, du = \pi$ either by numerical or contour integration.

(b) Show that $\displaystyle\int_{-\infty}^{\infty} f(a) \left(\int_{-q}^{q} e^{iat}\, dt \right) da = \int_{-\infty}^{\infty} f(a) \frac{2\sin aq}{aq}\, q\, da$.

(c) Substitute $u = aq$ and let $q \to \infty$ to show that

$$\int_{-\infty}^{\infty} f(a) \left(\int_{-q}^{q} e^{iat}\, dt \right) da = 2 \int_{-\infty}^{\infty} f\!\left(\frac{u}{q}\right) \frac{\sin u}{u}\, du \to 2\pi f(0).$$

(d) Use the results of parts (a–c) to conclude that the integral representation of the Dirac delta function given in Equation 2.80 is correct.

(2.12) The discrete analog to the Dirac delta function is the Kronecker delta function $\delta_{k,k'}$, where k and k' are integer indices. The Kronecker delta function is discussed in more detail in Example 4.3, but briefly, $\delta_{k,k'}$ has the value of 1 if $k = k'$ and is zero otherwise. Consider a discrete summation representation:

$$\delta_{k,k'} = \frac{1}{N} \sum_{J=0}^{N-1} e^{\pm 2\pi i J(k-k')/N}, \quad (k-k') = 0, 1, 2, \ldots, N-1. \tag{2.98}$$

(a) Verify that Equation 2.98 is true when $k = k'$ for any value of $N = 1, 2, 3 \ldots$

(b) Let $z = e^{-2\pi i(k-k')/N}$ and use the geometric series

$$\sum_{J=0}^{N-1} z^J = \frac{1 - z^N}{1 - z}$$

to show that Equation 2.98 is also true when $k \neq k'$.

(c) Construct phasor diagrams for $(k - k') = 0$ and 1 with $N = 4$ to illustrate pictorially why Equation 2.98 is true.

(d) Show that Equation 2.80 can be reexpressed as $\displaystyle\delta(k) = \int_{-\infty}^{\infty} e^{\pm i2\pi ku}\, du$.

Comment on the similarities and differences between this equation and Equation 2.98.

(2.13) Use the following steps to show that the integral representation of the Dirac delta function can be used to verify Fourier integral theorem expressed in Equations 2.81 and 2.82.

(a) Replace the integration variable ω in Equation 2.81 by the dummy variable ω'.

(b) Substitute the revised version of Equation 2.81 for $G(t)$ into Equation 2.82. Explain why using a dummy variable of integration like ω' (instead of ω) is necessary to avoid confusion.

(c) Exchange the order of integration and perform the integration over t.

(d) Related to Example 2.1, look up Fourier integral theorem and state the conditions on a function $g(\omega)$ for its Fourier transform to exist. Find an example of function whose Fourier transform does not exist.

REFERENCES

Gamelin TW. 2003. *Complex Analysis*. New York: Springer.

Ghiglia DC and Pritt MD. 1998. *Two-Dimensional Phase Unwrapping: Theory, Algorithms, and Software*. New York: Wiley-Interscience.

Peierls R. 1979. *Surprises in Theoretical Physics*. Princeton, NJ: Princeton University Press.

Reif F. 1965. *Fundamentals of Statistical and Thermal Physics*. New York: McGraw-Hill.

FURTHER READING

http://www.math.vanderbilt.edu/~schectex/commerrs/

This website describes a wide variety of commonly committed mathematical errors.

Dirac PAM. 1958. *The Principles of Quantum Mechanics*, 4th ed. Oxford: Clarendon Press.

A description of delta functions and their applications to quantum mechanics is provided in this book.

3

LIMITING AND SPECIAL CASES

Equations and identities can be remarkably general in their scope. Often, the general form of an equation contains many interesting special cases, from which much useful information can be extracted. As described in Section 3.1, such cases can be used for the simple, but important, task of checking whether a general expression is correct. Alternatively, if the precise form of an equation or identity is still to be determined, then sometimes it can be found by investigating known special cases, as described in Section 3.2. This heuristic method is a powerful tool for practical problem solving. In Section 3.3, a method for finding a limit for the solution of a differential equation is presented. The chapter concludes with a discussion of transition points in equations and some associated physical phenomena, including critical points and phase transitions.

3.1 SPECIAL CASES TO SIMPLIFY AND CHECK ALGEBRA

Consider the well-known mathematical identity

$$\sin^2\theta + \cos^2\theta = 1. \tag{3.1}$$

Given any angle θ, the operation described on the left-hand side of Equation 3.1 *always* results in the value of 1. Because of our familiarity with Equation 3.1, we might take its generality for granted. Often, we can exploit

---

Thinking About Equations: A Practical Guide for Developing Mathematical Intuition in the Physical Sciences and Engineering, by Matt A. Bernstein and William A. Friedman

the generality of mathematical identities such as Equation 3.1 to simplify mathematical expressions.

For example, consider a partial fraction expansion

$$\frac{2x^2-5}{(x-2)(2x+1)(x-3)} = \frac{a}{(x-2)} + \frac{b}{(2x+1)} + \frac{c}{(x-3)}, \tag{3.2}$$

where a, b, and c are the coefficients to be determined. Partial fraction expansions like Equation 3.2 are useful because the right-hand side simplifies the calculation of integrals, Laplace transforms, and related mathematical operations.

There are a variety of methods to solve for a, b, and c, but we will focus on a particular procedure, sometimes called the "cover-up" method, which illustrates the advantage of carefully chosen special cases. Multiplying both sides of Equation 3.2 by $(x-2)(2x+1)(x-3)$ yields

$$2x^2-5 = a(2x+1)(x-3) + b(x-2)(x-3) + c(x-2)(2x+1). \tag{3.3}$$

Because we are assuming that Equation 3.3 holds for any value of x, we are free to choose specific values of x that simplify the job of finding the unknown coefficients. By setting $x=2$ in Equation 3.3, we eliminate two of the three terms from the right-hand side and readily find that $a = -3/5$. We can then repeat the procedure with $x = -1/2$ and $x = 3$, in each case eliminating terms to obtain the values of $b = -18/35$ and $c = 13/7$. Exploiting these special cases allows us to avoid a considerable amount of work.

The examination of special cases is also very useful for checking our work. It is easy to make an algebraic error when simplifying a complicated expression. Even when symbolic manipulation software is used, transcription errors can occur. Substitution of a few numerical values provides a very effective check for possible errors. Rederiving the same expression does not necessarily provide the same level of validation as a numerical check, because as humans, we are prone to repeating our errors.

As an example, consider a specific set of trigonometric expressions for quantities p and q that arise when we study the strength of a signal produced in a magnetic resonance imaging scanner

$$p \equiv 1 - a\cos\theta - b^2(a-\cos\theta),$$
$$q \equiv b(1-a)(1+\cos\theta), \tag{3.4}$$

where $0 < a < 1$ and $0 < b < 1$. Using the definitions in Equation 3.4, we can derive a useful identity

$$\frac{1-b^2}{\sqrt{p^2-q^2}} = \sqrt{\frac{1-b^2}{(1-a\cos\theta)^2 - b^2(a-\cos\theta)^2}}. \tag{3.5}$$

The derivation of Equation 3.5 is messy and is potentially fraught with sign errors and errors gathering terms, among others. We can gain confidence that Equation 3.5 is consistent with the definitions in Equation 3.4 by numerical substitution. Substituting a set of random values for a, b, and θ into the right-hand side of Equation 3.4 yields the corresponding values for p and q, which are then used to check Equation 3.5. Repeating this procedure with a new set of random values further increases our confidence. A computer program can rapidly verify a great number of cases, which provides strong empirical evidence that the equation is true.

We know that Equation 3.1 is correct (see Chapter 7, exercise 7.4), but suppose we want to check it numerically. It is straightforward to write a computer program to test whether

$$\left|\sin^2\theta + \cos^2\theta - 1\right| < \varepsilon \tag{3.6}$$

for a large number (e.g., 10^7) of random values of θ, where ε is some small tolerance (e.g., 10^{-8}) that accounts for round-off error in the numerical calculations.

Keep in mind that validation by numerical substitution of special cases does not constitute a mathematical proof of an identity. Such numerical validation is related to the topic of the next section.

3.2   SPECIAL CASES AND HEURISTIC ARGUMENTS

Ideally, we would derive all of our equations starting from first principles, but that is not always possible. In practice, we might only know the form of an equation for a few special cases or a limited range of the variables. These cases might be associated with an extreme limit (e.g., 0 or ∞) of a variable, or some other special value that simplifies the problem considerably. Fortunately, we usually can extract a considerable amount of information about the general solution by examining a few specific cases. We can rule out a proposed, but incorrect, general equation if we can show that it is inconsistent with even a single special case. In some instances, the special case might even give us a valuable hint and lead us toward the correct general equation.

Techniques that involve guessing at the solution are examples of *heuristic* reasoning. Heuristics is a problem-solving method that helps establish the plausibility of a result, especially by using trial-and-error methods. George Polya, a twentieth-century Hungarian and American mathematician, wrote extensively on heuristic and inductive reasoning. Not only mathematicians but also scientists and engineers are wise to heed a statement from his classic book *How to Solve It* (Polya 1957): "Heuristic reasoning is good in itself. What is bad is to mix up heuristic reasoning with rigorous proof. What is worse is to sell heuristic reasoning for rigorous proof."

We know from basic logic that if A implies B, and A is true, then B is also true. A central principle of heuristic reasoning is if A implies B, then our *confidence* in the truth of A increases if we can verify that B is true.

Suppose B is true and is a special case of a proposed general result A, and we are trying to verify A. We can progressively increase our confidence that our guess for the general result A is correct by finding additional, special cases B', B'', B''' ... , which are also true and are implied by A. If, however, we find that even one of the B', B'', B''' is false, then we must conclude that A is not true in general (see exercise 3.8).

We illustrate these ideas with Example 3.1, which is based on a simple geometric construction. We then discuss blackbody radiation, where analogous methods were historically used to make an important scientific discovery. We return to a related topic in Chapter 7, where we further discuss techniques to generalize equations.

EXAMPLE 3.1

Consider a square with sides of length L. Inside of this square, we construct another square, as illustrated by the shaded region on Fig. 3.1. The inner square is constructed by drawing four line segments, which are shown as dashed lines. Each line segment makes the same angle θ with the nearest side of the outer square. We seek a mathematical expression for the area A of the inner central

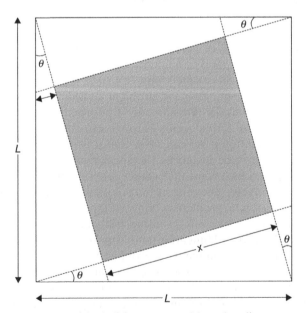

FIG. 3.1. A central square (shaded) is constructed from four line segments (dashed).

square as a function of L and θ. We only consider the range $0 \le \theta \le 45° = \pi/4$ rad, because each larger value of θ up to $90°$ replicates a configuration from the smaller angular range. Parts (a–e) of the example illustrate a heuristic line of reasoning for obtaining the desired mathematical expression for the area.

(a) What is the area A of the central square when $\theta = 0$? $\theta = 45°$?

(b) Based on dimensional considerations and on the specific values described in (a), write down several trial equations that could possibly describe the area A as a function of L and θ.

(c) Based on Fig. 3.2, determine, to first order in θ, the dependence of the area A for small values of the angle. Use the results of this limiting case to rule out some of the trial equations proposed in part (b).

(d) Determine the area for the specific case suggested by Fig. 3.3. Use this result to further rule out possible equations for the area A.

(e) Determine a general expression for the area of the central square directly using trigonometry. Compare the result with the remaining equation(s) after the incorrect ones were eliminated in parts (c) and (d).

ANSWER

(a) When $\theta = 0$, the central square coincides with the outer square, so the shaded area $A = L^2$. When $\theta = 45°$, all of the dashed lines meet together at

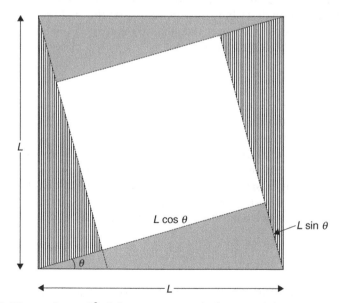

FIG. 3.2. The total area L^2 of the outer square is the sum of the area of the central square and four right triangles (shaded).

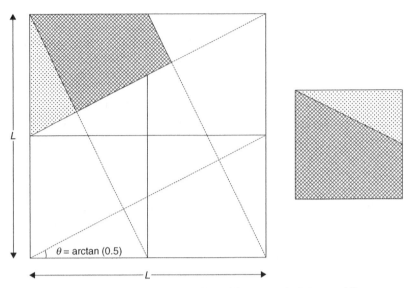

FIG. 3.3. A special case considered in Example 3.1, part (d).

the center of the square. Consequently, the central square shrinks to a point, and $A = 0$.

(b) We can obtain further information about the area A without knowing the general equation describing its dependence on the angle θ. From dimensional considerations, we expect the general expression for the area of the central square to be of the form

$$A = L^2 \times g(\theta), \tag{3.7}$$

where the function g is to be determined. We know that g must be dimensionless, because the factor L^2 provides the dimensions required for the area. Also, because there is only one quantity L with the dimensions of length given in the problem, we can infer that $g(\theta)$ cannot contain any dependence on L. If the problem had involved two quantities with the dimensions of length, say L_1 and L_2 instead of the single quantity L, then dimensional considerations would have allowed the general form of the solution to take a more complicated form such as

$$A = L_1^a \times L_2^b \times G\left(\theta, \frac{L_1}{L_2}\right), \tag{3.8}$$

with $a + b = 2$. Given only the single quantity L that has dimensions of length, however, the answer must follow the simpler form of Equation 3.7. Dimensional analysis is discussed in more detail in Chapter 6.

Because area cannot be negative, the value of g cannot be negative for $0 \leq \theta \leq 45°$. Finally, Fig. 3.1 also suggests that the area A monotonically decreases as θ increases from 0 to $45°$.

From parts (a) and (b), we know the value of function g at two specific values of θ: $g(0) = 1$ and $g(45°) = 0$. There are many possible functions that produce these two specific values. When trying to guess the general solution, we typically start with simple trial functions that fit the known cases. We hope that the desired solution is among these guesses, but naturally, there is no guarantee. Possible choices include

$$A_1(\theta) = L^2 \cos 2\theta,$$
$$A_2(\theta) = L^2 (1 - \tan \theta),$$
$$A_3(\theta) = L^2 (1 - \sin 2\theta), \tag{3.9}$$
$$A_4(\theta) = L^2 \left(1 - \frac{4\theta}{\pi}\right),$$

which seem plausible, and all satisfy the constraints considered so far. Because Fig. 3.1 contains many right triangles, the linear function A_4 might seem less likely, but including it in Equation 3.9 cannot do any harm.

Many other functions, however, besides those listed in Equation 3.9 work equally well. The list of the functions can be expanded by generalizing the expressions in Equation 3.9. Using the fact that for any positive exponent x, $0^x = 0$ and $1^x = 1$, the list of candidates becomes

$$A_5(\theta) = L^2 \cos^n 2\theta$$
$$A_6(\theta) = L^2 (1 - \tan^n \theta)^m$$
$$A_7(\theta) = L^2 (1 - \sin^n 2\theta)^m \qquad m > 0, n > 0 \cdot \tag{3.10}$$
$$A_8(\theta) = L^2 \left(1 - \left(\frac{4\theta}{\pi}\right)^n\right)^m$$

Because Equation 3.10 includes many possible functions, more information is needed to eliminate some of them.

(c) The behavior of the area for $0 < \theta \ll 1$ rad provides further insight into the acceptable solutions. As depicted in Fig. 3.2, the total area of the outer square L^2 consists of the unknown area A, plus the area of the four shaded triangles. It is straightforward to show that the height of each shaded triangle is approximately $L \times \theta$ for small values of θ. Because the base is length L, the area of each of the shaded triangles is approximately $L^2\theta/2$. Geometrically, this result follows from the observation that, for sufficiently small values of θ, each of the shaded triangles

TABLE 3.1. Dependence of the Trial Function in Equation 3.10 for $\theta \ll 1$ Rad.

Trial Equation	Leading Term in θ
$A_5(\theta) = L^2 \cos^n 2\theta$	$L^2(1 - 2n\theta^2 + ...)$
$A_6(\theta) = L^2(1 - \tan^n \theta)^m$	$L^2(1 - m\theta^n + ...)$
$A_7(\theta) = L^2(1 - \sin^n 2\theta)^m$	$L^2(1 - m(2\theta)^n + ...)$
$A_8(\theta) = L^2\left(1 - \left(\dfrac{4\theta}{\pi}\right)^n\right)^m$	$L^2\left(1 - m\left(\dfrac{4\theta}{\pi}\right)^n + ...\right)$

becomes indistinguishable from a circular wedge. Consequently, to first order in θ

$$A \approx L^2 - 4 \times \left(\frac{1}{2}L^2\theta\right) = L^2(1 - 2\theta), \quad \theta \ll 1 \, \text{rad}. \tag{3.11}$$

Table 3.1 shows the small-angle expansions of the various trial expressions proposed in Equation 3.10. Based on comparison with Equation 3.11, we can eliminate A_5 because the leading term is quadratic in θ. Similarly, A_6 is consistent with the limiting case only when $n = 1$ and $m = 2$. The function A_7 works only if $n = 1$ and $m = 1$. Finally, the small-angle limiting case requires that $n = 1$ and $m = \pi/2$ for A_8. The limiting case described by Equation 3.11 has ruled out many of the possible solutions, but many remain. Further information is needed to find the most likely solution.

(d) Knowledge of the value of the area A at a third specific value of θ allows further narrowing of the possible choices. One such specific case is illustrated in Fig. 3.3: $\theta = \arctan(0.5)$ or $\theta \approx 26.565°$. At first, this seems like an unlikely value to choose for the angle, but it does have the property that each dashed line exactly bisects one side of the outer square. Another special case is investigated in exercise 3.3.

From the geometric construction suggested by Fig. 3.3, the entire area of the outer square is covered by five smaller squares of equal area, one of which is the central square that has area A. Consequently, we can conclude that

$$A\left(\arctan\frac{1}{2}\right) = \frac{L^2}{5}. \tag{3.12}$$

Because $1 - \sin(2 \times \arctan 0.5)$ does indeed equal $1/5$, we can be reasonably confident that the function A_3 from Equation 3.9 is the correct answer. None of the other candidates satisfies Equation 3.12. Any other function that satisfies all of the necessary constraints is likely to have a much more complicated form. Our confidence in Equation 3.9 is further bolstered by another principle of heuristic reasoning known as Occam's razor, which can be paraphrased by saying the simplest explanation that fits the known data is likely to be the correct solution.

(e) There are several ways to solve for A directly by applying trigonometry, i.e., by applying first principles, rather than by guessing. Referring back to Fig. 3.2, we can show that the area of each of the four shaded triangles is $(L^2/2)\sin\theta\cos\theta$. By applying the trigonometric identity $\sin 2\theta = 2\sin\theta\cos\theta$, the result $A = A_3 = L^2(1 - \sin 2\theta)$ follows directly.

One could certainly argue that finding the area for the special case $\theta = \arctan(1/2)$ in part (d) of Example 3.1 is more difficult than deriving the general solution in part (e)! Many of the example problems presented in this book are idealized to illustrate a broader concept, and admittedly, Example 3.1 is somewhat contrived. But it does illustrate an important and practical point. In many real-life problems that arise in science and engineering, a closed-form solution developed from first principles is difficult or impossible to obtain. We might not even completely understand the basic principles required to set up the equations. We still can make progress by constructing a mathematical model represented by an equation that is consistent with all of the known special cases. In these real-life problems, each special case might be obtained from an experiment, or perhaps a numerical simulation.

Heuristic methods were used to advance the study of blackbody radiation, which played an important role in the development of modern physics at the turn of the twentieth century. A "blackbody" is an object that absorbs all light (i.e., electromagnetic radiation), both visible and invisible. A lump of graphite fits this criterion reasonably well. A blackbody also radiates light, and the properties of the emitted spectrum depend on the blackbody's temperature. At temperatures normally encountered in day-to-day life, the blackbody radiates light predominately in the infrared range, which is invisible to us. As the temperature is increased above approximately 1000K, however, it begins to glow in the visible range.

In 1896, the German physicist Wilhelm Wien published an equation that agreed with previous measurements for the amount of light energy u radiated by a blackbody (per unit volume and per unit frequency). This quantity is expressed in terms of the frequency f (measured in hertz) and the absolute temperature T (measured in kelvins) in what became known as Wien's distribution or Wien's approximation:

$$u_W(f, T) = \frac{\alpha f^3}{e^{(\beta f/T)}}. \tag{3.13}$$

The constants α and β can be determined from the measured spectral data.

Wien's law, however, did not agree with subsequent measurements for the energy radiated at lower values of the frequency f. In 1900, another German physicist, Max Planck, proposed his famous revision to Equation 3.13:

$$u_P(f, T) = \frac{8\pi f^2}{c^3} \times \frac{hf}{e^{\frac{hf}{kT}} - 1}. \tag{3.14}$$

Planck succeeded in specifying the empirical constants in Wien's law in terms of fundamental constants, i.e., $\alpha = 8\pi h/c^3$ and $\beta = h/k$, where c is the speed of light in meters per second, k is Boltzmann's constant in joules per kelvin, and h is a constant (measured in joule-seconds) that later became known as Planck's constant. The available measurements for blackbody spectra were sufficiently accurate for Planck to determine the value of h to within approximately 1% of the value accepted today.

The introduction of the "−1" term in the denominator of Equation 3.14 is quite important. When the dimensionless quantity (hf/kT) is large, the exponential dominates that constant term, so that Equation 3.14 reduces to Wien's distribution. For lower frequencies, however, the presence of the "−1" term allows the expression to differ greatly from Equation 3.13 and to fit the measured low-frequency data as well.

A limiting case was also used to develop the specific form of Equation 3.14. We have written Equation 3.14 as the product of two factors because each one has its own physical meaning. The denominator of the second factor can be expanded using $e^x = 1 + x^2/2 + \ldots$ to give

$$\frac{hf}{e^{\frac{hf}{kT}} - 1} = kT\left(1 - \frac{hf}{2kT} + \ldots\right), \quad hf \ll kT. \tag{3.15}$$

This factor represents the average energy per mode, where a "mode" can be thought of as a distinct solution to the wave equation for light. From the work of his contemporaries in the field of statistical mechanics, Planck knew that for small values of hf/kT, the average energy per mode should reduce to kT, in agreement with Equation 3.15. This limiting case, along with the specific values provided by experimental measurements of blackbody radiation, was the basis for Equation 3.14.

Ironically, historical accounts suggest that Planck was dissatisfied with Equation 3.14, because its derivation was based on heuristic reasoning, rather than on first principles. A derivation of Equation 3.14 from first principles was not available until further insight was gained into the nature of light over the next 25 years by Planck's famous countryman Albert Einstein and Satyendra Nath Bose, an Indian mathematician and physicist. Still, Equation 3.14 accurately models the measured data for blackbody radiation and introduced an important new fundamental constant h. As such, the introduction of Equation 3.14 proved instrumental in the further development of modern physics.

3.3 LIMITING CASES OF A DIFFERENTIAL EQUATION

So far in this chapter, we have seen how special and limiting cases are useful for a wide range of tasks ranging from checking algebraic steps to the development of an important scientific theory. Many additional applications of special and limiting cases are neither so mundane nor so ambitious, and we draw an example from the field of ordinary differential equations.

A mathematical function of a variable can be specified in several different ways. Ideally, we have an explicit formula, into which we can input values of the independent variable and generate a table of values of the function. This table can be graphed to visualize the function and to help interpolate between discrete points. Alternatively, if we do not have an explicit formula, we might know a series expansion, such as

$$\cos x = 1 - \frac{x^2}{2!} + \frac{x^4}{4!} + \dots, \tag{3.16}$$

which can be quite accurate over some range of the variable. For example, if we retain up to the fourth-power term in x in Equation 3.16, then the expansion is accurate to within 0.3% for $|x| < 1$ rad.

Yet another way to specify a function is by providing a differential equation that it satisfies, along with a set of boundary conditions. Often, the boundary conditions are specified at a particular value of the independent variable, such as the time we start an experiment, $t = 0$. In that case, the boundary conditions are called "initial conditions." The differential equation together with its initial conditions comprise an "initial value problem." A theorem described in standard texts on differential equations states that, *provided* certain conditions are satisfied, the solution to the initial value problem is unique. (See exercise 3.6 for further discussion and an example illustrating what can go awry when the conditions are not met.) For example, the function $y = \cos(x)$ uniquely satisfies a second-order, linear differential equation

$$\frac{d^2 y}{dx^2} + y = 0 \tag{3.17}$$

when the set of initial conditions $y(0) = 1$ and $y'(0) = 0$ are specified at $x = 0$. If we are given only the differential equation, but not the initial conditions, then more than one function can solve the equation. In the case of Equation 3.17, $y = \sin(x)$ is also a valid solution corresponding to the initial conditions $y(0) = 0$ and $y'(0) = 1$. In the absence of specified boundary conditions, then the linear combination

$$A \cos x + B \sin x \qquad\qquad (3.18)$$

is a general solution to Equation 3.17, where the arbitrary constants A and B can be determined when the boundary conditions are specified.

Differential equations are classified into various categories. Equation 3.17 is said to be an ordinary, second-order, linear differential equation. It is "ordinary" because y is a function of only the single variable x, and no partial derivatives appear. It is second-order because y'' (the second derivative of y with respect to x) is the highest derivative that appears. Equation 3.17 is linear because y is raised to the first power wherever it appears. Because no term contains a function of x or a constant alone without an accompanying factor of y, Equation 3.17 is also said to be *homogeneous*.

Certain differential equations arise repeatedly in the study of mathematics, physical science, and engineering. Many of these differential equations have been investigated thoroughly, and their solutions have been well documented in standard references such as Abramowitz and Stegun (1965). Because these differential equations recur frequently, many of their solutions are given dedicated symbols and names and are called "special" functions. A few examples of these functions include Bessel functions, Legendre polynomials, and the gamma function.

Occasionally, we do not need to know the general form of the special function, but instead, we only want to know a limiting case, that is, its values or functional form as the argument grows either very large (toward $-\infty$ or $+\infty$) or very small (toward zero).

If we have an explicit formula for the function, the task is straightforward. We can find the limiting case by examining the form of the function or perhaps by evaluating it for a few values in the desired range. A series expansion can also be useful for determining limiting cases when x approaches zero, e.g., we can see from Equation 3.16 that $\cos x$ approaches 1 as $x \to 0$. Next, we will discuss an alternative approach for finding the limiting form of a function when it is specified by a differential equation, rather than by an explicit formula or a power series. Using this method, we can find a functional form that provides a good approximation to the exact function in the desired limit. To illustrate these ideas, we will consider two examples of special functions: the well-known spherical Bessel function and the somewhat lesser-known Struve function.

Some problems, particularly those having intrinsic spherical symmetry, give rise to differential equations involving the Laplacian ∇^2 of a function, which can be solved in terms of a set of special functions called the spherical Bessel functions. These functions are named in honor of the German mathematician and astronomer Friedrich Bessel, who first described them in the early nineteenth century. Bessel also described an even better known, related set of special functions denoted $J_n(x)$ and $Y_n(x)$, which are useful for problems with cylindrical symmetry.

A spherical Bessel function of order n satisfies the second-order, linear, homogeneous differential equation

$$r^2 \frac{d^2 y}{dr^2} + 2r \frac{dy}{dr} + \left(r^2 - n(n+1)\right) y = 0, \tag{3.19}$$

where we will take n to be a nonnegative integer, $n = 0, 1, 2, \dots$ and $r \geq 0$. Quite analogous to what we saw with Equation 3.17 with its sine and cosine solutions, there are independent types of spherical Bessel functions, each satisfying a different set of initial conditions. The type of spherical Bessel function that is regular (i.e., does not approach $-\infty$ or $+\infty$) at $x = 0$ is called the spherical Bessel function of the first kind and is denoted $j_n(r)$.

We will omit the derivation here, but given the appropriate boundary conditions for the solution of Equation 3.19, the spherical Bessel functions of the first kind can be expressed in terms of the familiar sine and cosine functions and inverse powers of r,

$$j_0(r) = \frac{\sin r}{r},$$

$$j_1(r) = \left(\frac{\sin r}{r^2} - \frac{\cos r}{r}\right), \tag{3.20}$$

and so on for higher orders of n. We will use the known form for $j_1(r)$ given in Equation 3.20 to test the procedure for finding limits from the differential equation. Then, we will apply the same procedure to another special function, called the Struve function of order 1/2, which is denoted $\mathbf{H}_{1/2}(r)$. This function is named after Karl Hermann Struve, a nineteenth-century German astronomer. The Struve function $y = \mathbf{H}_{1/2}(r)$ satisfies the second-order, linear differential equation

$$r^2 \frac{d^2 y}{dx^2} + r \frac{dy}{dx} + \left(r^2 - \frac{1}{4}\right) y = \sqrt{\frac{2}{\pi}} r^{3/2}. \tag{3.21}$$

Note that Equation 3.21 is *not* homogeneous because the term on the right-hand side contains a function of r alone, with no accompanying factor of y.

EXAMPLE 3.2

(a) Find the r-dependence of the spherical Bessel function $j_1(r)$ for small values of r by examining the differential equation given in Equation 3.19.

(b) Alternatively, find the r-dependence of $j_1(r)$ for small values of r directly from the explicit formulas given in Equation 3.20.

(c) Find the r-dependence of $j_1(r)$ for large values of r by examining Equation 3.19.

(d) Find the behavior of the Struve function $\mathbf{H}_{1/2}(r)$ for small values of r by examining Equation 3.21.

ANSWER

(a) The spherical Bessel function $j_1(r)$ satisfies Equation 3.19, which for $n = 1$ reduces to

$$r^2\frac{d^2y}{dr^2} + 2r\frac{dy}{dr} + (r^2 - 2)y = 0. \tag{3.22}$$

If we are interested in values of y for $r \ll \sqrt{2}$ (i.e., r is much less than $\sqrt{2}$), we will neglect r^2 in the last term to obtain

$$r^2\frac{d^2y}{dr^2} + 2r\frac{dy}{dr} - 2y = 0. \tag{3.23}$$

Equation 3.23 is a progressively more accurate approximation to Equation 3.22 as $r \to 0$. Substituting the trial solution $y = cr^a$ into Equation 3.23 yields

$$(a(a-1) + 2a - 2) \times cr^a = 0, \tag{3.24}$$

which provides the values $a = 1$ and $a = -2$. The value $a = 1$ is associated with the spherical Bessel function of the first kind $j_1(r)$, which is regular at $r = 0$. The value $a = -2$ is associated with another solution to the same differential equation that approaches $\pm\infty$ at the origin. It is called the spherical Bessel function of the second kind and is usually denoted $y_1(r)$.

We draw the conclusion that the regular solution has the limit $j_1(r) \to cr^1$. With this procedure, we can only determine the limiting form up to a multiplicative constant c. This is to be expected. The expression in Equation 3.23 is homogeneous in y, so multiplying any valid solution by a constant results in another valid solution.

(b) From the specific form of $j_1(r)$ in Equation 3.20 and with the use of the standard series expansions of sine and cosine (e.g., see Equation 3.16), we obtain

$$j_1(r) \to \frac{1}{r^2}\left(r - \frac{r^3}{3!} + \ldots\right) - \frac{1}{r}\left(1 - \frac{r^2}{2!} + \ldots\right) = \frac{r}{3} + \ldots \tag{3.25}$$

For the solution to be consistent with Equation 3.20, the unknown multiplicative constant in part (a) is $c = 1/3$.

The small r limit of the solution can be extended to include more terms, i.e., terms proportional to r^2, r^3, etc., by an iterative procedure that is outlined in Chapter 5, exercise 5.10, which generates a series that agrees with Equation 3.25.

(c) We can also use Equation 3.22 to examine the behavior of $j_1(r)$ for large values of r. If $r \gg \sqrt{2}$, we will neglect the "2" in the last term of Equation 3.22 and examine the equation

$$r^2 \frac{d^2 y}{dr^2} + 2r \frac{dy}{dr} + r^2 y = 0. \tag{3.26}$$

By making the variable substitution $y = z/r$ into Equation 3.26 and working through the derivatives and algebra, we eventually obtain a simpler equation for z:

$$\frac{d^2 z}{dr^2} + z = 0. \tag{3.27}$$

Equation 3.27 has the same form as Equation 3.17, so the general solution to Equation 3.27 is

$$z(r) = A \cos r + B \sin r. \tag{3.28}$$

Therefore, we expect that

$$j_1(r) \rightarrow \frac{A \cos r}{r} + \frac{B \sin r}{r} \tag{3.29}$$

for $r \gg \sqrt{2}$. We cannot determine the arbitrary constants A and B solely from Equation 3.26 because no boundary conditions have been specified. If we return to the explicit form for $j_1(r)$ shown in Equation 3.20 and examine its dependence for large values of r, we can infer that

$$A = -1, \quad B = 0 \tag{3.30}$$

Boundary conditions of the spherical Bessel functions are further explored in exercise 3.5.

(d) We now use the technique developed in part (a) to find the small r behavior of the Struve function $\mathbf{H}_{1/2}(r)$. We are interested in values of $r \ll 1/2$, so we will neglect r^2 in the last term of Equation 3.21, yielding

$$r^2 \frac{d^2 y}{dr^2} + r \frac{dy}{dr} - \frac{y}{4} = \sqrt{\frac{2}{\pi}} r^{3/2}. \tag{3.31}$$

We will apply the trial solution $y = kr^b$ to Equation 3.31, which gives

$$(b(b-1)+b)kr^b - \frac{1}{4}kr^b = \sqrt{\frac{2}{\pi}} r^{3/2}. \tag{3.32}$$

We can infer from the exponents in Equation 3.32 that $b = 3/2$. Substituting $b = 3/2$ yields

$$\left[k \left(\frac{3}{4} + \frac{3}{2} - \frac{1}{4} \right) - \sqrt{\frac{2}{\pi}} \right] r^{3/2} = 0, \tag{3.33}$$

which is true when $k = 1/\sqrt{2\pi}$. So the behavior for small r of the order-1/2 Struve function is

$$\mathbf{H}_{1/2}(r) \rightarrow \frac{r^{3/2}}{\sqrt{2\pi}}. \tag{3.34}$$

Notice that in this case, unlike as in part (a), we obtained an answer which includes the specific value of the multiplicative constant. This is because Equation 3.31 is not homogeneous in y. As in part (a), however, we obtained the desired result without directly solving the complete differential equation (i.e., Equation 3.21 or 3.22), so we again avoided a lot of work.

We have not proved that for any arbitrary differential equation, the solution of a particular limiting case must equal the limiting case of the solution. In the case of the Bessel and Struve functions described in Example 3.2, this switching of the orders of these two operations is valid, as can be verified with numerical checks or an iterative procedure described in exercise 5.9 in Chapter 5.

3.4 TRANSITION POINTS

The number of real solutions to an equation can depend on the specific values of the parameters appearing in it. A familiar example is the quadratic equation

$$ax^2 + bx + c = 0, \tag{3.35}$$

where the coefficients are real numbers and $a \neq 0$. The well-known solution to Equation 3.35 is

$$x = \frac{-b \pm \sqrt{b^2 - 4ac}}{2a}, \qquad (3.36)$$

from which we can see that Equation 3.35 has zero, one, or two real roots depending on the combination of parameters known as the discriminant, $d = b^2 - 4ac$. To illustrate this result graphically, we recast Equation 3.35 or 3.36 into the form

$$\left(x + \frac{b}{2a}\right)^2 = \frac{d}{4a^2} \qquad (3.37)$$

and then plot each side of Equation 3.37 as a function of x. The left-hand side gives a parabola, and the right-hand side gives a horizontal line. Any point of intersection between the two plots represents a real solution to Equation 3.35. Graphical solutions to equations are explored further in Chapter 4.

Figure 3.4 illustrates the result represented by Equation 3.36 that when $d > 0$ there are two real solutions, when $d = 0$ there is a single real solution at $x = -b/(2a)$, and when $d < 0$ there are no real solutions. The value $d = 0$ represents a *transition point* for the number of solutions to Equation 3.37.

This familiar dependence of the number of real solutions to Equation 3.35 on the discriminant d serves as a prototype for the behavior of many equations that model physical systems. In those equations, a change in the number of real solutions often can be linked to interesting, observable behavior. For example, the emergence or disappearance of a solution occurs at a *critical point**, where it is directly related to an observable physical phenomenon.

Figure 3.5 shows a schematic diagram depicting a critical point that is often studied in the field of thermodynamics. The plot shows the values pressure P and temperature T, at which the liquid and gaseous phases of a substance (like water and steam) can coexist in a stable state. At each of the points along the plotted curve, a *phase transition* occurs, that is, the liquid boils and the gas condenses. Beyond the critical point, no such phase transition occurs. Later in this section, we will describe how the coexistence of the two phases mathematically corresponds to two physical solutions of an equation. At temperatures and pressures beyond the critical point, however, there is only one physical solution to that equation, and the substance can only exist in a single phase. These effects are analogous to the simple example of the number of real roots to the quadratic equation. We begin with another example of a phase transition, the effect of temperature on a permanent magnet.

---

*In mathematics, the term "critical point" has a slightly different meaning. There, it is understood to be a maximum, minimum, or inflection point of a function, i.e., a point where the first derivative is zero. We will see from the analysis of the van der Waals equation of state how the two meanings can be related.

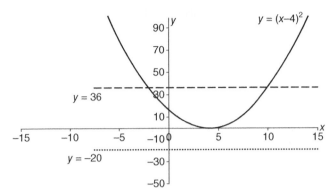

FIG. 3.4. Real roots to the quadratic equation can be found from a graphical construction based on Equation 3.37.

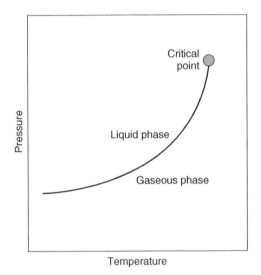

FIG. 3.5. A schematic plot showing a phase diagram for a liquid–gas system, such as water and steam. For each point along the solid curve, the liquid and gas phases can coexist. Beyond the critical point, however, the system exists only in a single phase.

Some metallic elements including iron, cobalt, and nickel, as well as many alloys such as steel, can be permanently magnetized and are said to be ferromagnetic. When a ferromagnetic material is magnetized by placing it in a strong external magnetic field, the directions of a large number of microscopic magnetic moment vectors become aligned with each other. This alignment persists even after the external field is switched off. Experiments show, however, that the metal cannot retain this magnetization if its temperature exceeds a critical value called its Curie point or *Curie temperature* T_C, named after Pierre Curie, a nineteenth-century French physicist. Measurements show that the value of the Curie temperature varies among elements and alloys. In iron, the Curie temperature is $T_C = 1043\,\text{K}$, or $770\,°\text{C}$.

The "mean-field" approach is a simple model that helps us understand ferromagnetism and the Curie temperature. In this approach, we assume that every atomic magnetic moment experiences a uniform magnetic field B that is self-consistent with the effects of all the other surrounding magnetic moments. The mathematics that results is reminiscent of the graphic solution quadratic equation, which is illustrated in Fig. 3.4.

Under the influence of a uniform magnetic field B and absolute temperature T (in kelvins), we can use statistical physics to calculate the average probability that the atomic magnetic moments align in the direction of B. We will omit the derivation here, but the result is that the average magnetization M (i.e., magnetic moment per unit volume) is given by the function

$$M = n\mu \tanh\left(\frac{\mu B}{kT}\right), \tag{3.38}$$

where μ is the magnetic moment of an individual atom, n is the number of atoms per unit volume, and k is Boltzmann's constant. Notice that the argument of the hyperbolic tangent (a transcendental function) is dimensionless because both the numerator and denominator have the same dimensions, namely energy. The dimension of M is magnetic moment per volume.

We make Equation 3.38 self-consistent by assuming that the magnetic field B is in turn proportional to the average magnetization M, with proportionality constant λ:

$$B = \lambda M. \tag{3.39}$$

Using Equation 3.39 to eliminate the magnetic field B from Equation 3.38 yields a single equation involving M:

$$M = n\mu \tanh\left(\frac{\mu \lambda M}{kT}\right). \tag{3.40}$$

We define the Curie temperature in terms of parameters appearing in Equation 3.40:

$$T_C \equiv \frac{n\lambda\mu^2}{k}. \tag{3.41}$$

We also define the dimensionless magnetization parameter:

$$m \equiv \frac{M}{n\mu}. \tag{3.42}$$

Then, Equation 3.40 is expressed more compactly:

$$m = \tanh\left(\frac{mT_C}{T}\right). \tag{3.43}$$

The value of the hyperbolic tangent varies from 0 to 1 as its argument ranges from 0 to ∞. Equation 3.43 tells us that as the temperature T approaches absolute zero, the argument of the hyperbolic tangent goes to ∞, so the value of m approaches 1. Physically, this means that the atomic magnetic moments all become aligned with each other.

EXAMPLE 3.3

Show that $m = 0$ is a solution to Equation 3.43. Show that a second real solution for m emerges only when $T < T_C$.

ANSWER

Because $\tanh(0) = 0$, $m = 0$ is always a solution to Equation 3.43. Given a specific value for the temperature ratio (T/T_C), we can use numerical methods to search for other real solutions of Equation 3.43. A graphic solution can be obtained by plotting the left-hand side $y = m$ and the right-hand side $y = \tanh(mT_C/T)$ as a function of m and by looking for points of intersection. These points of intersection correspond to the real solutions.

Figure 3.6 shows plots of $y = \tanh(mT_C/T)$ versus m for three different values of T: $0.5T_C$, T_C, and $2T_C$. Above the Curie temperature, e.g., for $T = 2T_C$, the sole real solution is $m = 0$. Therefore, the mean-field model predicts that we cannot sustain permanent magnetism in the metal when the temperature is that high. For $T < T_C$, however, there is a real, positive solution. For example, when $T/T_C = 0.5$, we find that $m = 0.9575$ is a solution.

The behavior of Equation 3.43 is interesting near $m = 0$, but it is somewhat difficult to determine from the graph, which is cluttered near the origin. The series expansion

$$\tanh x = x - \frac{1}{3}x^3 + \ldots, \quad |x| < \frac{\pi}{2} \tag{3.44}$$

helps us verify that when $T \geq T_C$, the plot of the hyperbolic tangent remains below the identity line $y = m$, and thus the only intersection point occurs at $m = 0$. Equation 3.44 also shows this for small values of mT_C/T:

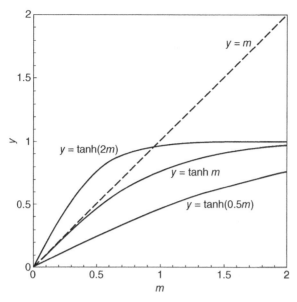

FIG. 3.6. A plot of Equation 3.43 illustrating the effect of temperature on a ferromagnetic material. Points of intersection with the dashed line indicate real solutions.

$$\tanh\left(\frac{mT_C}{T}\right) \approx m \times \frac{T_C}{T}. \tag{3.45}$$

Below the Curie temperature, the ratio of temperatures is greater than 1, so Equation 3.45 ensures that $\tanh(mT_C/T) > m$. The hyperbolic tangent asymptotically approaches the value of 1 as its argument approaches ∞. Consequently for $T < T_C$, there is an additional solution for m with $0 < m \leq 1$. The value of m at the point of intersection (i.e., the solution) increases toward its final value of 1 as the temperature T decreases toward zero.

Another well-known example of critical phenomena is related to the equation of state for gases and liquids that was described by the Dutch scientist Johannes Diderik van der Waals in his doctoral thesis. That equation can be used to describe the transitions between the gaseous and liquid phases of a substance, i.e., condensation and vaporization. This has practical applications when studying how to liquefy helium or determining the temperature and pressure at which water boils. Van der Waals received the Nobel Prize in physics in 1910 for his work.

In thermodynamics, a confined gas with a fixed number of particles N is characterized by its pressure P, volume V, and temperature T. Writing the pressure P as some function f of V and T provides an *equation of state*:

$$P = f(V, T). \tag{3.46}$$

For example, the well-known ideal gas law

$$PV = NkT \tag{3.47}$$

describes a gas of noninteracting particles (i.e., atoms or molecules). Because $V > 0$, Equation 3.47 is equivalent to the equation of state

$$P(V, T) = \frac{NkT}{V}. \tag{3.48}$$

When there are interactions between the gas particles, however, the situation is more complicated than Equation 3.48 can model. An attractive interaction can cause condensation of the gas into a liquid. Van der Waals modified the equation of state in Equation 3.48 to account for an attractive force between the particles at large separations and for a repulsive force at small separations:

$$P(V, T) = \frac{NkT}{(V - Nb)} - \frac{N^2 a}{V^2}. \tag{3.49}$$

The parameter a has dimensions of force $\times$ distance^4 and is a measure of the attractive force between two particles. The parameter b has the dimension of volume and reflects the nonzero volume occupied by the particles themselves. Typically, parameters a and b are fit to experimental data, although b can be estimated if the molecular dimensions are known. Notice that when a and b are both zero, Equation 3.49 reduces to the ideal gas law, Equation 3.48.

EXAMPLE 3.4

Using Equation 3.49, plot curves of pressure versus volume for various values of temperature. Show that there can be up to three values of volume V that correspond to a specific value of the pressure P for some values of T, while for other values of T, show there is only a single solution for V.

ANSWER

Figure 3.7 shows a plot of Equation 3.49 for three different *isotherms*, that is, plots of pressure versus volume when the temperature is held fixed. The isotherms are plotted for $T = 0.9 T_{\text{crit}}$, T_{crit}, and $1.1 T_{\text{crit}}$ where

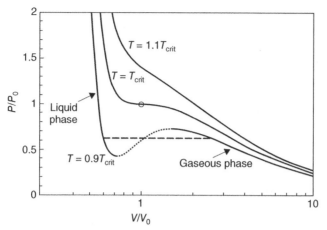

FIG. 3.7. A plot of pressure versus volume as described in Equation 3.49 for three different values of the temperature. The small circle indicates a critical point.

$$T_{crit} = \frac{8a}{27bk}. \tag{3.50}$$

As discussed in exercise 3.7, the plot of $P(V, T)$ versus V contains an inflection point on the isotherm $T = T_{crit}$, i.e., at that point both the first and second partial derivatives of P with respect to V are zero. (Recall that an inflection point is a "critical point," when the term is used in its mathematical sense.) At the inflection point, the values of pressure and volume are

$$P_{crit} = \frac{a}{27b^2} \quad \text{and} \quad V_{crit} = 3Nb. \tag{3.51}$$

As a specific example, experiments show that the critical isotherm for carbon dioxide occurs at $T_{crit} = 31\,°C$, and the inflection point occurs at $V_{crit} = 1.3 \times 10^{-4}\,m^3$ and $P_{crit} = 7370$ kPa, or approximately 74 times atmospheric pressure.

From Fig. 3.7, we see that for $T > T_{crit}$, there is a one-to-one relationship between the variables P and V, and the substance exists as a single phase. For $T < T_{crit}$, however, there is a minimum and a maximum in $P(V)$. In this range of temperatures, the equation of state suggests that there can be up to three values of the volume V (i.e., solutions) corresponding to a single value of the pressure P. The critical isotherm $T = T_{crit}$ represents a transition in the number of solutions. At lower temperatures, there can be high-volume, low-volume, and intermediate-volume solutions.

The van der Waals equation of state describes many physical situations, but additional physical constraints rule out some of the solutions, i.e., intersection points of an isotherm with the horizontal line $P = const$. For example, along the dotted line in Fig. 3.7, the pressure increases while the volume also increases. That dependence is not mechanically stable, so any solution corresponding to a system associated with those intermediate values of V cannot be achieved physically and can be ignored.

In the study of thermodynamics, we deal with several quantities known collectively as *free energy*. These quantities represent the energy content in a closed system that can be converted into work. We omit the details here, but the specific type of free energy depends on the conditions under which the system does the work. For example, at constant T and P, we have the Gibbs free energy G, while at constant T and V, we consider the Helmholtz free energy A. The determination of the free energy specifically excludes the contribution of energy from the random thermal motion of the molecules (the entropy term), because that energy cannot be harnessed. It also excludes any additional energy required to hold non-varying parameters constant while the system does work. The important point for our discussion is at a particular temperature and pressure, a physical system is stable only in a configuration where the Gibbs free energy is a minimum. That constraint allows us to rule out some additional sections along the isotherms described by the van der Waals equation of state.

If the system is colder than the critical temperature T_{crit}, it can be shown that for almost all values of the pressure P in Fig. 3.7, the free energies are unequal at the two remaining solutions. (Recall that we excluded the mechanically unstable, third solution.) That means the system exists either in its low-volume phase (liquid) or its high-volume phase (gas), whichever has the lower value of the free energy. The system does not coexist in both phases simultaneously at those pressures.

At a particular pressure (aptly called the coexistence pressure), however, the liquid and gaseous phases have equal values of the free energy, so it is allowable for them to coexist. For the isotherm corresponding to $T = 0.9T_{crit}$ in Fig. 3.7, these two points correspond to the end points of the dashed line. This leads to the wide range of possible volumes, i.e., those along the dashed line, corresponding to varying fractions of the molecules occupying the liquid or gaseous phases.

This situation for the other isotherms (that have a coexistence pressure) is also indicated by the set of points that make up the solid curve in the schematic plot of Fig. 3.5, which ends in the critical point. Thus, in Fig. 3.5, the points along the curve indicate the value of the coexistence pressure at each value of the temperature. It is at the coexistence conditions of pressure and temperature that liquid boils and the gas condenses.

Suppose a system of molecules is held at a fixed temperature that is below T_{crit}. Also, suppose that the system is initially at high pressure so it occupies the liquid phase. As the pressure is decreased, the volume increases, and the system continues to be stable as a liquid until the coexistence pressure is reached, where the gaseous and liquid phases coexist. As the pressure is further decreased, the system is again stable but is now solely in the gaseous phase. The unstable region shown by the dotted line in Fig. 3.7 is avoided.

In our everyday experience, we are more familiar with boiling a liquid by increasing its temperature rather than by decreasing its pressure. Typically, the pressure is held fixed at whatever value the atmospheric pressure happens to be on that day. Figure 3.5 suggests that the temperature of the boiling point increases with temperature. This result should not be surprising to anyone who is familiar with the operation of a pressure cooker to prepare food.

The following story is told about how the elevation of the Dead Sea was discovered during the nineteenth century. Some Englishmen brought their native habit of drinking tea to the region. They noticed that the water took an unusually long time to boil and that their tea seemed hotter than what they remembered experiencing back in England. Eventually, they realized these observations were due to the unusually high atmospheric pressure at the shore of the Dead Sea. They further realized that the increased atmospheric pressure was the result of their elevation being well below sea level. Using modern instruments, which were not available at that time, it was later determined that the shore of the Dead Sea is the lowest land on the surface of the Earth. It lies approximately 418 m (or more than a quarter of a mile) below sea level.

The change of ferromagnetic properties at the Curie temperature and the boiling of a liquid are two prototypical examples of phase transitions. In each case, the number and location of real roots to an equation provide insight into physical processes that we can observe in the lab. These two examples of the phase transitions can be linked, or unified, with a theory of phase transitions developed by the twentieth-century Russian physicist Lev Davidovic Landau.

If we consider a system with a fixed number of particles N, then the Landau theory of phase transitions deals with the free energy. In the discussion of the van der Waals equation of state, we saw that the constraint that the free energy be a minimum at equilibrium plays an important role in determining the conditions under which the system is stable.

The Landau theory of phase transitions, in its simplest schematic form, introduces the general parameter x, which is called the *order parameter* (and is sometimes also denoted by the Greek letter ξ). The order parameter x along with the temperature T are the arguments of the expression for the Landau free energy $F(T, x)$. The parameter x can represent a different quantity in each specific physical system to be examined, but it is invariably associated with some degree of order, with $x = 0$ corresponding to a disordered state. For the case of ferromagnetism, we will take x to be the normalized magnetization m, as defined by Equation 3.42.

For our purposes, it is sufficient to use x to examine the general mathematical behavior of the free energy function, $F(T, x)$. We gain insight into phase transitions by exploring the location of the minima of F with respect to x over certain ranges of the temperature parameter T. These minima are located by finding the real roots of the equation

$$\frac{\partial F(T, x)}{\partial x} = 0, \tag{3.52}$$

subject to the condition that the second derivative is positive:

$$\frac{\partial^2 F(T, x)}{\partial x^2} > 0. \tag{3.53}$$

The real values of x that satisfy Equations 3.52 and 3.53 are minima, and the global minimum is associated with a stable state of the system. By monitoring the number of solutions as a function of T, we can determine the temperature at which a phase transition will occur.

We illustrate these ideas by examining two applications of the Landau theory. First, to model the ferromagnetic transition using the Landau model, we assume that the free energy can be approximated as a series of even terms of the order parameter x:

$$F(T, x) = F_0(T) + \frac{1}{2}\alpha(T - T_C)x^2 + \frac{1}{4}g_4 x^4. \tag{3.54}$$

The rationale behind the construction of Equation 3.54 is the $(T - T_C)$ factor ensures that the quadratic term vanishes at the Curie temperature. The expression is constructed exclusively of even-order terms because we would expect the free energy not to change if the direction of the magnetization is inverted (i.e., $m \rightarrow -m$), provided there is no externally applied magnetic field. Equation 3.54 is expected to be valid only for $T \approx T_C$, where the order parameter x is small, so that only low-order powers are needed. The form of Equation 3.54 is sufficient, however, to study the properties of the phase transition.

EXAMPLE 3.5

Find the minimum of the Landau free energy given in Equation 3.54. Consider the cases (a) $T \geq T_C$ and (b) $T < T_C$.

ANSWER

The minima of Equation 3.54 are found by solving

$$\frac{\partial F}{\partial x} = (T - T_C)\alpha x + g_4 x^3 = 0, \tag{3.55}$$

subject to the constraint that the second derivative is positive.

(a) For $T \geq T_C$, the sole minimum occurs at $x = 0$.

(b) For $T < T_C$, the Landau free energy has a minimum at

$$x = \sqrt{\frac{\alpha(T_C - T)}{g_4}}, \qquad (3.56)$$

Considered as a function of increasing T, the value of the order parameter x in Equation 3.56 decreases to zero when $T = T_C$. For higher values of T, the minimum occurs at $x = 0$. Consequently, the location of the minimum is a continuous function of x. From Equation 3.56, however, the derivative dx/dT is discontinuous. This type of behavior is a characteristic of a *second-order phase transition*. Ferromagnetism is an example of a second-order phase transition because the value of m is continuous at the Curie temperature, but the derivative dm/dT is discontinuous there.

The Landau model can also be applied to the van der Waals equation of state for the liquid–gas system. In this case, we model the free energy as

$$F(T, x) = F_0(T) + \frac{1}{2}\alpha(T - T_0)x^2 - \frac{1}{4}g_4 x^4 + \frac{1}{6}g_6 x^6, \qquad (3.57)$$

where α, g_4, and g_6 are positive constants. Note that the sign of the coefficient of the x^4 term has been chosen to be negative in Equation 3.57. That term provides a dip in F at intermediate values of x. The addition of the positive sixth-order term overwhelms the effect of the fourth-order term for larger values of x, so these two terms work in concert to provide a minimum.

EXAMPLE 3.6

Find the minima in Equation 3.57 for:

(a) $T = T_0$,
(b) $T < T_0$, and
(c) $T > T_0$.
(d) Show that at a particular value of the temperature, there are two minima that have equal values of the Landau free energy.

ANSWER

We find that minimum by solving for the real roots of

$$\frac{\partial F}{\partial x} = (T - T_0)\alpha x - g_4 x^3 + g_6 x^5 = 0. \tag{3.58}$$

Equation 3.58 always has one solution at $x = 0$, which is a minimum for $T > T_0$. If $x \neq 0$, we can cancel one power of x to obtain a quadratic equation in x^2. The solution is

$$x_{min}^2 = \frac{g_4 \pm \sqrt{g_4^2 - 4g_6\alpha(T - T_0)}}{2g_6}. \tag{3.59}$$

(a) At $T = T_0$, Equation 3.59 reduces to

$$x_{min,0} = \sqrt{\frac{g_4}{g_6}}, \tag{3.60}$$

which yields

$$F(T_0, x_{min,0}) = F_0(T_0) - \frac{1}{12}\frac{g_4^3}{g_6^2}. \tag{3.61}$$

The free energy in Equation 3.61 is less than its value at $x = 0$. Therefore, $x_{min,0}$ represents a global minimum and a stable phase of the system at $T = T_0$.

(b) For $T < T_0$, there is also a single deep minimum at $x_{min}^2 > g_4/g_6$ that is also the global minimum.

(c) Above the temperature $T > T_0$, there are two interesting regions. We can see from Equation 3.59 that the argument of the square root becomes negative when

$$T > T_0 + \frac{g_4^2}{4\alpha g_6}. \tag{3.62}$$

So at higher temperatures, i.e., values of T for which the inequality (Eq. 3.62) is satisfied, the sole minimum occurs at $x = 0$.

(d) In the intermediate region,

$$T_0 < T < T_0 + \frac{g_4^2}{4\alpha g_6}, \tag{3.63}$$

we look for a value of T where

$$F(x = 0, T_t) = F(x_{min,t}, T_t). \tag{3.64}$$

When Equation 3.64 is satisfied, we have the coexistence situation that allows the system to make a phase transition back and forth from the disordered state characterized by $x = 0$ to the more ordered state with $x_{min,t} > 0$. With the use of Equations 3.59 and 3.57 and some algebra, we can show that Equation 3.64 is satisfied when

$$T_t = \left(T_0 + \frac{3}{16} \frac{g_4^2}{\alpha g_6} \right). \tag{3.65}$$

We can now see that the temperature parameter T_0 introduced in Equation 3.57 is not the transition temperature. We then can show that when $T = T_t$, Equation 3.64 is satisfied when

$$x_{min,t} = \sqrt{\frac{3g_4}{4g_6}}. \tag{3.66}$$

Comparing Equations 3.60 and 3.66, we find that the minimum of $F(T, x)$ decreases from $\sqrt{g_4/g_6}$ to $\sqrt{0.75g_4/g_6}$ as T increases from T_0 to T_t. At T_t, the two states of the system coexist, and a phase transition such as the liquid boiling occurs. Then at higher temperatures, $x = 0$ is the global minimum and represents the sole equilibrium state of the system. Figure 3.8 shows plots of $F(x, T)$ versus x for $T = T_0$, T_t, and $T_0 + 0.375g_4^2/(g_6\alpha)$ with $g_4 = 4$ J and $g_6 = 6$ J, illustrating these three cases. (The plots are valid for any positive value of α, which has the unit of joules per kelvin.)

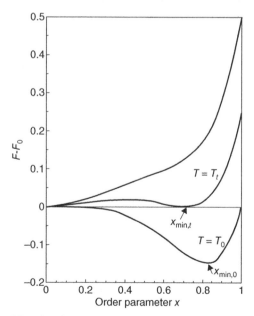

FIG. 3.8. A plot of Landau free energy versus the order parameter x, as described in Equation 3.57. The curve labeled $T = T_t$ has two equal minima in the free energy, indicating that the system is allowed to make a phase transition.

Phase transition occurs at T_t, which is known as the *transition temperature*. Notice that when $g_4 > 0$, the second solution $x_{min,t}$ is not contiguous with the solution at $x = 0$. This discontinuous behavior in the order parameter x characterizes a *first-order phase transition*. Physically, latent heat energy must be added to or released by a system undergoing a first-order phase transition. This effect is familiar to us from our everyday experience with boiling water or observing steam condense. For the ferromagnetic (i.e., second order) phase transition discussed earlier, however, no latent heat energy is involved because there is no jump in the order parameter x.

For the van der Waals liquid–gas, the critical temperature T_{crit} is higher than the transition temperature T_t. We can include this dependence in the model by modifying the equations. In Equation 3.57, the overall sign of the fourth-order term is always negative. We could have instead modeled the g_4 coefficient itself to be a function of temperature, for example, $g_4 = \beta(T_{crit} - T)$, where β is a positive constant. Then, for $T > T_{crit}$ the overall sign of the fourth-order becomes positive, and there is no minimum in the Landau free energy, except at $x = 0$. At lower temperatures, the overall sign is negative, and the previous cases are recovered. Notice that for $T = T_{crit}$, g_4 goes to zero, so the two coexisting values of x involved in the phase transition merge because they are separated by $\sqrt{0.75 g_4 / g_6}$.

The phase transitions illustrated in Examples 3.3–3.6 serve as prototypical examples, but transition points also occur across a wide variety of equations. As illustrated in exercise 3.4, even a simple sign change in a variable can represent interesting changes in observable behavior. The emergence or disappearance of real roots also occurs in many equations not related to thermodynamic phase transitions. This is illustrated in Chapter 8, Example 8.2, where we analyze the situation of a rope draped over two pegs. The problem then concerns finding the length of the rope for which equilibrium can be maintained, given the configuration of peg positions. Depending on the positions of the pegs, there can be zero, one, or two real solutions for the length of the rope that correspond to a stable configuration.

EXERCISES

(3.1) Consider the partial fractions expansion

$$\frac{3x^2 - 12x + 2}{(x+1)(x-2)^3} = \frac{A}{x+1} + \frac{B}{(x-2)^3} + \frac{C}{(x-2)^2} + \frac{D}{x-2}, \tag{3.67}$$

where A, B, C, and D are coefficients to be found.

(a) Multiply both sides of Equation 3.67 by $(x + 1)$ and set $x = -1$ to find A.

(b) Multiply both sides of Equation 3.67 by x and let $x \to \infty$ to show that $A + D = 0$, thereby determining D.

(c) Use appropriate special cases for x to find the remaining coefficients, B and C.

(3.2) Suppose you derived the trigonometric identity

$$\tan 5\theta = \frac{\tan^5 \theta - 10\tan^3 \theta + 5\tan \theta}{1 - 10\tan^2 \theta + 5\tan^4 \theta} \tag{3.68}$$

and want to verify that you did not make a sign error or some other algebraic error.

(a) Use the method suggested by Equation 3.6 to check the identity for three random values of θ.

(b) Examine Equation 3.68 for any particular values of θ that could be problematic, such as values that cause the denominator to go to zero, or $\tan \theta \to \infty$.

(3.3) Using Fig. 3.9, find another special case related to Example 3.1. Show that when $\theta = \arctan (3/4)$, the area of the central square is $L^2/25$.

(3.4) A *sign change* is another type of transition point in an equation that can indicate a dramatic change in physical behavior. Consider the spool of

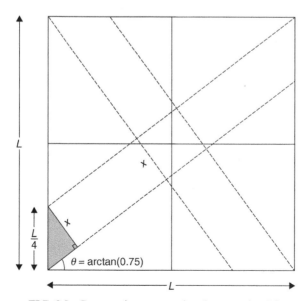

FIG. 3.9. Geometric construction for exercise 3.3.

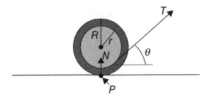

FIG. 3.10. A spool of thread as described in exercise 3.4.

thread resting on a level surface, as schematically shown in Fig. 3.10. The spool has mass M and moment of inertial I and is lying on a horizontal surface with rolling frictional force f. The outer radius of the spool is R, and the radius of the coiled thread is r. The following analysis can predict whether the spool rolls toward or away from us as we pull the thread with tension T so that the thread makes an angle $0 \le \theta \le 90°$ with the horizontal.

Let the normal force of the surface on the spool be N. If the spool rolls, then there is no vertical acceleration, and the vertical forces balance:

$$N = mg - T\sin\theta. \tag{3.69}$$

Let a be the horizontal acceleration of the spool. Our main goal is to determine the sign of a. We will adopt a convention that positive values of a mean that the spool accelerates in the direction of the pulling force. Then Newton's second law gives

$$F\cos\theta - f = Ma, \tag{3.70}$$

where f is the frictional force.

If the spool rolls without slipping, then its angular acceleration α is related to the linear acceleration by

$$\alpha = \frac{a}{R} \tag{3.71}$$

and satisfies the equation

$$fR - Fr = I\alpha. \tag{3.72}$$

(a) Using Equations 3.69–3.72, eliminate N, α, f, and g to solve for the acceleration a:

$$a = \frac{F\left(\cos\theta - \dfrac{r}{R}\right)}{M + \dfrac{I}{R^2}}. \tag{3.73}$$

(b) Verify that Equation 3.73 is a dimensionally consistent equation.

(c) According to Equation 3.73, the acceleration undergoes a sign change when $\theta_c = \arccos(r/R)$. When $\theta < \theta_c$, pulling the thread rolls the spool toward us and winds up the spool. When $\theta > \theta_c$, pulling the thread causes the spool to unwind and accelerate away from us.

(d) Obtain a physical spool of thread from a sewing kit, a kite, etc. Replicate the experiment suggested by parts (a–c). Vary the angle $0 \leq \theta \leq 90°$. Observe which direction the spool rolls. What happens when $\theta = \theta_c$? (The experiment works better if there is substantial friction between the spool and the surface. Use of a heavier spool and a rubber surface such as on the face of a ping-pong paddle helps.)

(e) Returning to Fig. 3.10, show that when $\theta = \theta_c$, all of the external forces on the spool of thread pass through an axis containing the point P. That implies that the net torque about that axis is zero, and so the angular acceleration about point P is also zero. Why is this result consistent with Equations 3.71 and 3.73?

(3.5)

(a) Consider the spherical Bessel functions $j_0(r)$ and $j_1(r)$ given in Equation 3.20. Show that they satisfy Rayleigh's formula:

$$j_n(r) = r^n \left(-\frac{1}{r}\frac{d}{dr} \right)^n \frac{\sin r}{r}. \tag{3.74}$$

(b) Use Equation 3.74 to find $j_2(r)$.

(c) Show that as $r \to 0$, $j_0(r), j_1(r)$, and $j_2(r)$ satisfy the limit:

$$j_n(r) \to \frac{r^n}{(1 \times 3 \times 5 \times \ldots \times 2n+1)}. \tag{3.75}$$

If we want to, we can use Equation 3.75 to specify initial conditions for $j_n(r)|_{r=0}$ and $j'_n(r)|_{r=0}$ for the differential equation shown in Equation 3.19.

(d) The boundary conditions for the $j_n(r)$ alternatively can be specified at large values of r with the expression

$$j_n(r) \to \frac{\cos\left(r - \frac{(n+1)\pi}{2} \right)}{r}. \tag{3.76}$$

Show that at the explicit forms for $j_0(r), j_1(r)$ and $j_2(r)$ are consistent with Equation 3.76.

(e) Using the trigonometric identity for the cosine of a sum, show that Equation 3.76 is consistent with Equation 3.29 and that the values for A and B given in Equation 3.30 are correct.

(3.6) Consider the initial value problem

$$\frac{d^2y}{dt^2} - g = 0, \tag{3.77}$$

where $g = 9.8$ m/s^2 is the constant acceleration due to gravity, and the initial conditions are $y(0) = 0$ and $y'(0) = 0$. Physically, this initial value problem models an object being dropped from a height $y = 0$ at a time $t = 0$, and $y(t)$ is the distance that it falls after a time t. Equation 3.77 neglects the effects of air resistance.

(a) Show that

$$y = \frac{1}{2}gt^2 \tag{3.78}$$

is a solution to the initial value problem. Differentiate Equation 3.78 to obtain

$$\frac{dy}{dt} = gt. \tag{3.79}$$

(b) Use Equation 3.78 to eliminate t from the right-hand side of Equation 3.79 to obtain the first-order, nonlinear differential equation

$$\frac{dy}{dt} = \sqrt{2gy} \tag{3.80}$$

with the initial condition $y(0) = 0$.

(c) Show that Equation 3.78 is indeed a valid solution for the initial value problem described by Equation 3.80, but it is *not* a unique solution. Show that $y(t) = 0$ works as well. The solution $y(t) = 0$ is not physically meaningful; it corresponds to the object being released and then floating, unsuspended in air. Because Equation 3.80 has more than a single solution, it is not useful for modeling the physical problem of the free-falling object. As mentioned in Section 3.3, there is a uniqueness theorem for initial value problems, but clearly, Equation 3.80 is violating at least one of its conditions. As described in texts on differential equations, one of the conditions for the solution to be unique is that the derivative of the right-hand side of Equation 3.80,

$$\frac{d}{dy}\sqrt{2gy} = \sqrt{\frac{g}{2y}}, \tag{3.81}$$

must be bounded. From Equation 3.81, this condition is violated in the neighborhood of $y = 0$.

(3.7) Consider the van der Waals equation of state:

$$P(V, T) = \frac{NkT}{(V - Nb)} - \frac{N^2 a}{V^2}.$$

(a) Holding the temperature T constant, calculate the first and second partial derivative of P with respect to V to show that

$$\frac{\partial P}{\partial V} = \frac{-NkT}{(V - Nb)^2} + \frac{2an^2}{V^3},$$

$$\frac{\partial^2 P}{\partial V^2} = \frac{2NkT}{(V - Nb)^3} - \frac{6aN^2}{V^4}. \tag{3.82}$$

Show that Equation 3.82 and the conditions for an inflection point,

$$\frac{\partial P}{\partial V} = 0 \quad \text{and} \quad \frac{\partial^2 P}{\partial V^2} = 0,$$

can be solved to show that the critical points are

$$T_{crit} = \frac{8a}{27bk}, \quad P_{crit} = \frac{a}{27b^2}, \quad \text{and} \quad V_{crit} = 3Nb. \tag{3.83}$$

(b) Show that in terms of dimensionless variables, the van der Waals equation of state can be written simply as

$$p = \frac{8t}{3v - 1} - \frac{3}{v^2}, \tag{3.84}$$

where $p = P/P_{crit}$, $t = T/T_{crit}$, and $v = V/V_{crit}$. Note that the number of particles N and the empirical parameters a and b do not appear explicitly in Equation 3.84, so it is a universal relation that is true for any quantity or type of substance obeying van der Waals equation of state. The advantages of expressing equations in terms of dimensionless variables, as in Equation 3.84, are discussed further in Chapter 6.

(3.8) Consider the general proposition A: "all animals are primates." Find two special cases B and B' that increase our confidence that A could be true. Find another special case B'' that demonstrates that A is false. Discuss how the improper application of heuristic reasoning can lead to a faulty conclusion.

REFERENCES

Abramowitz M and Stegun IA. 1965. *Handbook of Mathematical Functions: With Formulas, Graphs, and Mathematical Tables*. New York: Dover.

Polya G. 1957. *How to Solve It, a New Aspect of the Mathematical Method*. Princeton, NJ: Princeton University Press.

FURTHER READING

Polya G. 1954. *Mathematics and Plausible Reasoning, Volume 1, Induction and Analogy in Mathematics*. Princeton, NJ: Princeton University Press.

A more in-depth treatment on heuristic methods is given here. Both this book and *How to Solve It* are highly recommended.

Tenenbaum M and Pollard H. 1985. *Ordinary Differential Equations*. New York: Dover Publications.

A thorough introductory treatment of ordinary differential equations, with many examples drawn from physics and engineering, is provided in this book.

Kittel C and Kroemer H. 1980. *Thermal Physics*, 2nd ed. New York: W.H. Freeman.

Much more background on the thermodynamics of phase transitions is provided in this book.

Sarhangi R, Jablan S, and Sazdanovic R. 2005. "Modularity in medieval Persian mosaics: Textual, empirical, analytical, and theoretical considerations." *Visual Mathematics Journal,* Vol. 7, No. 1.

This article explains how the tenth-century Persian mathematician and astronomer Abul Wafa al-Buzjani demonstrated a procedure for the dissection and reassembly of a large square into certain numbers of smaller squares of equal area, which relates to Figs. 3.3 and 3.9.

4

DIAGRAMS, GRAPHS, AND SYMMETRY

4.1 INTRODUCTION

Diagrams, graphs, and geometric constructions have always been closely related to equations. Many mathematicians and scientists have devoted great effort toward finding connections between geometric concepts and equations. Around 225 BC, the Greek mathematician, scientist, and engineer Archimedes determined a number of important geometric formulas, including the area of a circle in terms of its radius, which is a remarkable achievement given that he had no access to the modern tools of calculus. There are also connections between equations and diagrams, graphs, and geometry in many branches of modern mathematics. For example, graph theory deals with the study of pairwise relationships between objects. Graph theory makes extensive use of diagrams containing vertices and nodes and has practical applications such as finding efficient routes and managing traffic flow.

We are all familiar with the use of diagrams as an aid to performing everyday calculations. Consider a simple word problem such as:

An Olympic-sized swimming pool has dimensions $50 \times 25\,\mathrm{m}$. What is the total area of a 2-m-wide border that completely surrounds the pool?

A simple sketch helps us to arrive at the correct answer of $316\,\mathrm{m}^2$. Without the aid of a diagram, even a routine problem like this one can be challenging.

Thinking About Equations: A Practical Guide for Developing Mathematical Intuition in the Physical Sciences and Engineering, by Matt A. Bernstein and William A. Friedman
Copyright © 2009 John Wiley & Sons, Inc.

(Even if you do not draw a diagram explicitly to solve this simple problem, you probably at least visualize it.) The value of these diagrammatic aids seems to be universal. A recent scientific study by Dehaene et al. (2006) of an indigenous group in the Amazon provided evidence that the human brain is "hardwired" to respond to geometric concepts. That is, geometric intuition appears to be innate in all humans, regardless of educational experience.

This is not to say that every scientist, mathematician, and engineer has embraced the use of diagrams and figures. In the preface to his book *Mécanique Analytique*, the Italian-born, French mathematician Joseph-Louis Lagrange wrote, "The reader will find no figures in this work. The methods that I set forth do not require either constructions or geometrical or mechanical reasoning, but only algebraic operations, subject to a regular and uniform rule of procedure." For most of us, however, diagrams and figures greatly enhance our understanding of equations. In modern physics, the use of Feynman diagrams has become practically indispensable for modeling and understanding the interactions between particles. Figures and diagrams can accentuate the important features required to solve a problem, or suggest shortcuts.

In this chapter, we provide several examples that illustrate how diagrams, graphs, and other pictorial methods can help us understand equations. We also examine how diagrams can help us generate equations and sometimes even solve them. The chapter concludes with two examples that illustrate how symmetry arguments can greatly simplify solutions to problems.

4.2 DIAGRAMS FOR EQUATIONS

Certain equations are closely tied to a corresponding diagram. Consider the identity involving the sum of the first k odd integers:

$$1+3+5+\ldots+(2k-1)=k^2. \tag{4.1}$$

This identity can be readily verified using mathematical induction as follows. Equation 4.1 is true for the first term, i.e., $1 = 1^2$. Suppose it is also true for the first k terms. The next term on the left-hand side would be $(2k-1)+2=(2k+1)$, and adding this additional term gives the desired result:

$$[1+3+5+\ldots+(2k-1)]+(2k+1)=k^2+(2k+1)=(k+1)^2. \tag{4.2}$$

Instead of using this line of inductive reasoning, we can count squares in Fig. 4.1 to associate the odd, positive integers with blocks of area and thereby make a convincing argument for the validity of Equation 4.1. Diagrams such as Fig. 4.1 are sometimes called "proofs with pictures," or "proofs without words."

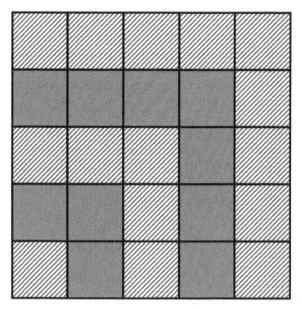

FIG. 4.1. Insight into the identity expressed in Equation 4.1 can be gained by counting squares for this case with $k = 5$.

While it is ultimately for mathematicians to decide whether or not methods like these meet the standard of a mathematical proof, most of us find these diagrams useful. We explore several more "proofs with pictures" in the exercises.

It is not always easy to generate a useful diagram to represent a mathematical relationship. In the case of the word problem about the area of the swimming pool border, drawing a useful diagram is easy, but in other cases like Equation 4.1, a little more thought is needed. In certain cases, the process is reversed, that is, a diagram suggests a key concept that is instrumental for generating the desired equation. This point is illustrated by the following example.

EXAMPLE 4.1

A well-known problem is to count the number of ways that objects can be distributed into containers. In the study of probability, this is usually called the "ball and urn" problem. We seek a count of the number of distinct ways that n balls can be distributed into r distinguishable urns. If the balls are also distinguishable from each other (like the numbered ping-pong balls used in a lottery game), the number of distinct configurations N is simply

$$N = r^n. \tag{4.3}$$

This result follows from the fact that the first ball can go into any of r urns, the second ball can independently go into any of r urns, and so on. So Equation 4.3 results from multiplying together one factor of r for each of the n balls. However, if the balls are *indistinguishable* from each other, then there is a fewer number N of distinct arrangements, and the resulting expression is more complicated than Equation 4.3.

(a) Find the number of ways that n indistinguishable balls can be distributed into r urns. Construct a diagram to make the analysis more apparent.

(b) Verify the general result found in (a) for the special cases $r = 1$ and $r = 2$ urns, and an arbitrary number of balls n. Then, verify the result for $n = 1$ ball and an arbitrary number of urns, r.

ANSWER

(a) As the values of n and r increase, it quickly becomes impractical to directly count all of the configurations. However, a useful diagram can be constructed based on the observation (i.e., the trick) that we can replace the r urns by $(r - 1)$ walls or partitions that separate them, as shown in Fig. 4.2. Then, we have a collection of both balls and partitions that has a total number of elements $n + (r - 1)$. We can think of these as $n + r - 1$ slots, each to be filled by one of the n balls or $r - 1$ partitions. An example with $n = 6$ and $r = 4$ yielding nine slots is shown in Fig. 4.3. With this diagram, our problem is equivalent to counting the number of ways to choose which of these $(n + r - 1)$ slots to fill with the n balls. The ordering of the n balls is immaterial, as is the ordering of the $r - 1$ partitions, so from the combinatorial analysis, the result is

$$N = \binom{n+r-1}{n} = \frac{(n+r-1)!}{n!\,(r-1)!}. \tag{4.4}$$

The binomial coefficient $\binom{a}{b} \equiv \frac{a!}{b!(a-b)!}, 0 \le b \le a$ is discussed in greater detail in Section 7.1.

(b) According to Equation 4.4, if $r = 1$, then $N = \binom{n}{n} = 1$. This result is clearly true because, with only one urn, all n balls must go into that one, so there is only a single configuration. If there are two urns, then Equation 4.4 yields

$$N = \binom{n+1}{1} = \frac{(n+1)!}{1!n!} = n+1. \tag{4.5}$$

n Balls

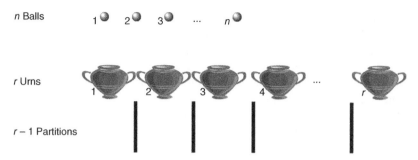

r Urns

r − 1 Partitions

FIG. 4.2. A classic problem from the study of probability is to count the number of ways that *n* balls can be distributed into *r* urns. The analysis can be simplified by replacing the *r* urns by *r* − 1 walls or partitions that separate the urns.

n = 6 balls

r = 4 urns

r − 1 = 3 partitions

n + r − 1 = 6 + 3 = 9 slots

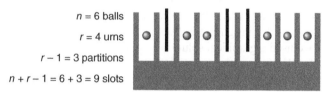

FIG. 4.3. Suppose there are six balls and four urns. This diagram depicts the particular configuration where there is one ball in the leftmost urn, two balls in the second urn, none in the third urn, and three balls in the last urn.

Equation 4.5 is also readily verified. Each of the *n* balls must go into either urn 1 or urn 2. Urn 1 can contain any number of balls from zero to *n*, with the remainder going into urn 2. There are *n* + 1 integers in the range [0, 1, 2, ... , *n*], in agreement with Equation 4.5.

Finally, if *n* = 1, then Equation 4.4 yields

$$N = \binom{r}{1} = r. \tag{4.6}$$

Equation 4.6 agrees with the expected result because the single ball can go into any of the *r* urns.

From Equation 4.4, we can begin to appreciate the futility of trying to solve this problem by directly constructing a different diagram for each possible arrangement. Even with modest numbers like *n* = *r* = 10, there are 92,378 distinct configurations. If we worked 24 hours per day and could draw two diagrams per minute, it would take over a month to complete the task. The small investment of time required to discover the method suggested by Fig. 4.2 is clearly worthwhile.

The result of Example 4.1 has several practical applications. For example, lasers operate by manipulating light particles, i.e., photons. In quantum optics, the photons in the laser are indistinguishable particles, i.e., Bose particles or "bosons." Results like Equation 4.4 allow us to count the number of ways that n photons can occupy r quantum states (analogous to the urns), which is useful for understanding the properties of a laser.

4.3 GRAPHICAL SOLUTIONS

Graphs and plots are often the most convenient way to organize and visualize the results of a numerical computation. With the widespread availability of computers today, the use of graphical methods to *solve* equations might seem less necessary than in the past. Although this is true to some extent, nonetheless graphical solutions still provide useful insight and effectively complement numerical methods. In Chapter 3, we used a graphical solution to help us understand ferromagnetism and the van der Waals equation of state. Even for a relatively simple numerical problem such as finding the real roots of the fifth-order polynomial,

$$f(x) = x^5 - x - 2 = 0, \tag{4.7}$$

we can plot $f(x)$ directly and look for points of intersection with the x-axis. The plot of Equation 4.7 can be readily displayed with a graphing calculator. If we rewrite Equation 4.7 as

$$x^5 = x + 2, \tag{4.8}$$

then we do not even need the calculator to determine the number of real roots and their approximate locations. We sketch the graphs of $y = x^5$ and $y = x + 2$ and see that there is only a single intersection point (near $x = 1.3$) that represents the sole real root of Equation 4.7. Having a rough idea about the location of the real root(s) is useful when setting the bounds on x for a more detailed numerical search. Notice that if the function in Equation 4.7 were $f(x) = x^5 - x$ instead, a sketch showing the points of intersection of $y = x^5$ and $y = x$ suggests that the number of real roots increases from 1 to 3.

In general, much insight can be gained about functional relationships with the simple use of graphs. As an even more basic example, consider the relationship between a function and its inverse. A function f, which relates y to x, i.e., $y = f(x)$, is plotted in Fig. 4.4. We can graphically find y by drawing a vertical line upward from any value x to the curve representing the function f, and then by proceeding horizontally from that intersection point to the value on the y-axis. This graphical construction is so basic that we tend to take it for granted. It is useful, however, for understanding why the inverse (if it exists) of a function f

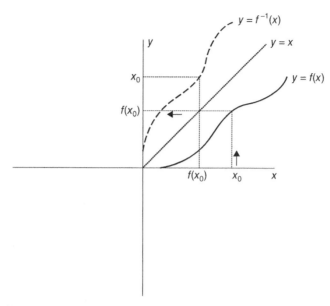

FIG. 4.4. The value of a function $y = f(x)$ at $x = x_0$ can be determined from its plot with the use of a simple graphical procedure. A related construction shows why the plot of the inverse function is reflected about the identity line $y = x$.

can be constructed by reflecting the plot of $y = f(x)$ about the identity line $y = x$. This graphical construction ensures that $x_0 = f^{-1}[f(x_0)]$, as is also shown in Fig. 4.4.

This basic example, which uses the graphs of a function and its inverse, has an extension that is important for an application in imaging science. Suppose that we wish to faithfully reproduce a paper image (i.e., make an exact copy), or instead make a reproduction with a different brightness, contrast, or some other attribute. A graphical process to accomplish these goals is illustrated next.

EXAMPLE 4.2

A simple monochrome, digital imaging system consists of a flatbed scanner, a computer, and a printer that are configured as shown in Fig. 4.5. The original paper image has a range of intensities characterized by the variable A_{in}, which depends on the location within the image. The value of A_{in} at a particular location on the paper image could represent the percent reflectance ranging from 0% to 100%, where the darker regions of the image have lower numerical values of A_{in} than the lighter regions.

The paper image is scanned, and the digital scanner subdivides the image into an array of picture elements or pixels. The output of the scanner is stored as a grayscale digital image on the computer with pixel intensities given by D_{in}. The digital image D_{in} could be stored as an array of pixel

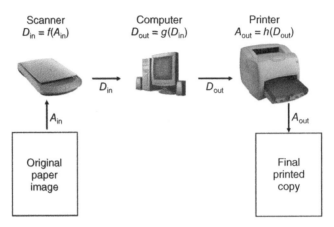

FIG. 4.5. A digital image reproduction system comprising a scanner, computer, and printer.

values represented by an 8-bit integer ranging from 0 to $2^8 - 1 = 255$. Mathematically, this digitization process can be described by the "tone transfer function" f, which is determined by the design and construction of the scanner:

$$D_{in} = f(A_{in}). \tag{4.9}$$

Likewise, the printer has a tone transfer function h, which is also determined by its own design and construction, and describes how light or dark various digital pixel values appear on the printed copy:

$$A_{out} = h(D_{out}). \tag{4.10}$$

The computer between the scanner and the printer contains a lookup table that relates the pixel values D_{in} to the corresponding values D_{out}. This lookup table is mathematically described by the function g, which is applied to each pixel in the digital image according to

$$D_{out} = g(D_{in}). \tag{4.11}$$

The function g provides the freedom to determine the final appearance of the printed image, which is determined by the entire transformation from A_{in} to A_{out}:

$$A_{out} = h(g(f(A_{in}))) \equiv e(A_{in}). \tag{4.12}$$

(a) Suppose that both the tone transfer functions f and h are known, or that they can be determined by a series of measurements on the scanner and printer, respectively. Write an expression that can be solved numerically for the function g (describing the lookup table) so that the intensity of the copied image exactly matches the original.

(b) Construct a graphical solution for the problem posed in (a). This is called a *Jones plot*, after Loyd A. Jones, an American physicist and imaging scientist.

ANSWER

(a) Equation 4.12 yields

$$g(f(A_{in})) = h^{-1}(e(A_{in})). \qquad (4.13)$$

In order for the copy to be a faithful reproduction, we require that $A_{out} = A_{in}$. This is equivalent to the function e for the entire imaging chain being the identity function, or

$$e(A_{in}) = A_{in}. \qquad (4.14)$$

Combining Equations 4.12 and 4.14 yields

$$A_{in} = A_{out} = h(g(f(A_{in}))). \qquad (4.15)$$

Therefore, for each value of A_{in}, the function g satisfies the relationship

$$g(f(A_{in})) = h^{-1}(A_{in}), \qquad (4.16)$$

where h^{-1} is the inverse of the function h. Given that f and h have been determined from measurements, Equation 4.16 can be solved for g numerically.

(b) The construction of the Jones plot is illustrated in Fig. 4.6. The two positive axes are labeled A_{out} and A_{in}, while the two negative axes are labeled D_{in} and D_{out}. Each quadrant is bounded by two axes, and the relationship between them is described by the functions e, h, g, and f for quadrants I–IV, respectively. The function that we want to construct graphically is the lookup table g. The functions f and h are assumed to be known, and the form of e is specified by our desired outcome, i.e., a straight diagonal line in the first quadrant representing the requirement that $A_{out} = A_{in}$.

For each point along the A_{in} axis, which represents an intensity value on the original image, we follow the graphical procedure illustrated in Fig. 4.6. This means that from point A_{in}, we work counterclockwise to A_{out} and then to D_{out}. Similarly, we work clockwise from A_{in} to D_{in}. This allows us to

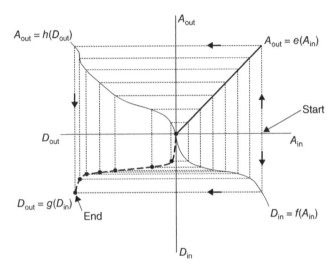

FIG. 4.6. A Jones plot is a graphical construction for the unknown function g. The identity line in the first quadrant indicates our desire for A_{out} to equal A_{in}.

construct one point on the graph of D_{in} versus D_{out}. Repeating the procedure for several different values of A_{in} constructs the entire graphical relationship between D_{in} and D_{out}, which yields the desired lookup table function g. (See exercise 4.3 for more insight into how the placement of the axes in the Jones plot allows for calculation of inverse functions, such as those that appear in Eq. 4.16.)

With the widespread availability of digital computers and numerical analysis software, the Jones plot is usually not needed to get numerical values for the lookup table g. It can provide, however, more insight than solving Equation 4.16 numerically and is still used. The plot gives a convenient overview of the entire imaging system and is especially instructive for suggesting procedures to handle troublesome regions that are not well suited to the use of numerical methods. Such regions might include places where the values of the measured functions f or h are uncertain due to noisy measurements, or missing altogether. A glance at the Jones plot will illustrate how these trouble areas in the measured data translate to errors or inaccuracies in g.

In some cases, we might want to print a copy that differs from the original. We might want it to be lighter or darker, or have more or less contrast. These changes can be accommodated by modifying the function e, that is, replacing the identity function in the first quadrant of the Jones plot, as explored in exercise 4.3.

Finally, we note that if the reader has ever downloaded a "printer driver" to a computer, that file probably contained a lookup table much like the function g.

Sometimes the opportunity to apply a graphical solution becomes available only midway through a calculation. In these cases, not only do we need to give some thought about which diagram or graph to construct but we also must recognize *when* it can be best introduced. This point is illustrated by the following example, which provides a graphical solution for finding eigenvalues* of a certain class of symmetric matrices.

EXAMPLE 4.3

Consider a 3×3 symmetric matrix of the form

$$\mathbf{M} = \begin{pmatrix} \varepsilon_1 + g_1^2 & g_1 g_2 & g_1 g_3 \\ g_1 g_2 & \varepsilon_2 + g_2^2 & g_2 g_3 \\ g_1 g_3 & g_2 g_3 & \varepsilon_3 + g_3^2 \end{pmatrix}. \tag{4.17}$$

Note that the off-diagonal matrix elements of $\mathbf{M}$ are the product of two factors $g_i g_j$.

A compact expression for the elements of an $N \times N$ matrix with this form is

$$M_{ij} = \varepsilon_i \delta_{ij} + g_i g_j, \quad i = 1, 2, \ldots, N, \quad j = 1, 2, \ldots, N, \tag{4.18}$$

where $\{g_i\}$ is a set of N numbers and δ_{ij} is the Kronecker delta, which was introduced in exercise 2.12. By convention, we represent the elements of the identity matrix $\mathbf{I}$ with the Kronecker delta δ_{ij}, which takes the value of 1 when $i = j$, but otherwise is zero. Consequently,

$$I_{ij} = \delta_{ij} = \begin{cases} 1, & i = j \\ 0, & i \neq j \end{cases}. \tag{4.19}$$

The $\{\varepsilon_i\}$ is also a set of N numbers, which range from the smallest value ε_{min} to the largest value ε_{max}.

To illustrate Equation 4.18 with a specific example, suppose that $N = 3$; $g_1 = \sqrt{3}$, $g_2 = \sqrt{4}$, and $g_3 = \sqrt{5}$; and $\varepsilon_1 = 1$, $\varepsilon_2 = 2$, and $\varepsilon_3 = 3$. Then clearly, $\varepsilon_{min} = 1$ and $\varepsilon_{max} = 3$. Using Equations 4.18 and 4.19, the 3×3 symmetric matrix for this specific case is

---

*The eigenvalues λ of a matrix $\mathbf{M}$ are discussed in standard texts on linear algebra. A reader who is unfamiliar with eigenvalues can also refer to Equation 4.21 for their definition and to Equation 4.36 for a standard method to calculate them.

$$\mathbf{M} = \begin{pmatrix} 1\times 1 & 0 & 0 \\ 0 & 1\times 2 & 0 \\ 0 & 0 & 1\times 3 \end{pmatrix} + \begin{pmatrix} 3 & \sqrt{12} & \sqrt{15} \\ \sqrt{12} & 4 & \sqrt{20} \\ \sqrt{15} & \sqrt{20} & 5 \end{pmatrix} \approx \begin{pmatrix} 4.000 & 3.464 & 3.873 \\ 3.464 & 6.000 & 4.472 \\ 3.873 & 4.472 & 8.000 \end{pmatrix}.$$

(4.20)

(a) Develop an equation that can be solved to find the N eigenvalues λ of the general matrix $\mathbf{M}$ given in Equation 4.18. (We could use a label a to distinguish among the various eigenvalues, i.e., λ^a, but to keep the notation simpler, we will suppress that superscript.)

(b) Plot the equation derived in (a) for the example matrix given in Equation 4.20 to find a graphical solution that illustrates the general behavior of the eigenvalues. Use this example to show that $(N-1)$ of the eigenvalues of $\mathbf{M}$ lie between ε_{min} and ε_{max}, while there is a single outlier.

(c) Suppose that all the values of ε_i are equal to a single value ε, and all of the values of g_i are equal to a single value g. Find the eigenvalues of $\mathbf{M}$ for this special case.

(d) Find the eigenvector corresponding to the outlying eigenvalue of $\mathbf{M}$ for the special case described in (c). Assume the eigenvectors are each normalized to 1, i.e., $\vec{v} \cdot \vec{v} = 1$.

(e) What happens to the eigenvalues of $\mathbf{M}$ if all of the values of g_i are purely imaginary numbers? That is, the products $(g_i g_j)$ that appear in Equation 4.18 are real, negative numbers.

ANSWER

(a) Our goal is to find the eigenvalues λ of the $N \times N$ matrix $\mathbf{M}$. The general equation to be solved is

$$\mathbf{M}\vec{v} = \lambda\vec{v} \qquad (4.21)$$

or, in terms of the matrix elements,

$$\sum_{j=0}^{N} M_{ij} v_j = \lambda v_i, \qquad (4.22)$$

where $\vec{v}$ is the eigenvector. Substituting Equation 4.18 into Equation 4.22, we find that

$$\sum_{j=0}^{N} (\varepsilon_i \delta_{ij} + g_i g_j) v_j = \lambda v_i. \qquad (4.23)$$

Using Equation 4.19 and noting that factor g_i can be brought outside the summation over the index j yields,

$$v_i(\varepsilon_i - \lambda) = -g_i \sum_{j=0}^{N} g_j v_j, \tag{4.24}$$

where we define the result of the summation on the right-hand side of Equation 4.24 to be C:

$$C = \sum_{j=0}^{N} g_j v_j. \tag{4.25}$$

Because each eigenvalue has its own eigenvector associated with it, each eigenvalue also corresponds to a specific value of C. (Again, we could use a label a to distinguish among the various eigenvalues, i.e., λ^a, v_j^a, C^a, but to keep the notation simpler, we are suppressing that superscript.)

Equation 4.24 can then be rearranged to give the elements of the eigenvector corresponding to each eigenvalue λ:

$$v_i = -\frac{g_i}{(\varepsilon_i - \lambda)} C. \tag{4.26}$$

Multiplying the right- and left-hand sides of Equation 4.26 by g_i and summing over i yields

$$\sum_{i=1}^{N} v_i g_i = C = -\sum_{i=1}^{N} \frac{g_i g_i}{(\varepsilon_i - \lambda)} C, \tag{4.27}$$

where we have used the fact that the sum in Equation 4.25 is equal to C regardless of whether the summation index is labeled i or j. Dividing Equation 4.27 by C and rearranging yields the desired result:

$$\sum_{i=1}^{N} \frac{g_i^2}{(\lambda - \varepsilon_i)} = 1. \tag{4.28}$$

Note that Equation 4.28 does not depend on the specific choice of the normalization of the eigenvectors, because neither C nor v appears. This is to be expected from Equation 4.21.

(b) The values of λ that satisfy Equation 4.28 can be found by plotting its left-hand side as a function of λ and observing where the graph crosses the value of 1. This is the point at which the graphical technique can be applied. For the specific example of the matrix in Equation 4.20, Equation 4.28 becomes

$$f(\lambda) = \frac{3}{\lambda - 1} + \frac{4}{\lambda - 2} + \frac{5}{\lambda - 3} = 1. \tag{4.29}$$

Figure 4.7 shows the graph of $f(\lambda)$. At each value of ε, $f(\lambda)$ diverges to $\pm\infty$ and undergoes a sign change. If $g_i^2 > 0$, as in our example, then each of

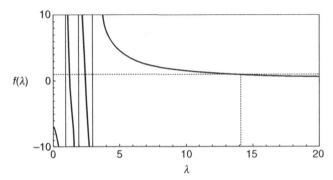

FIG. 4.7. A plot of Equation 4.29 provides insight into the eigenvalues of the matrix in Equation 4.20. We can infer useful features about the eigenvalues of the matrix in Equation 4.30 from this graphical solution.

the fractions in Equation 4.29 is negative for values of λ less than $\varepsilon_{min} = 1$. Therefore, the graph of $f(\lambda)$ cannot cross 1 in that region, so there are no eigenvalues less than ε_{min}.

We can also see from the graph that the equation $f(\lambda) = 1$ is satisfied exactly once between each pair of adjacent ε values. In this way, $N - 1$ of the eigenvalues are sandwiched between ε_{min} and ε_{max}. In our numerical example, this means that two of the three eigenvalues occur between 1 and 3. The largest eigenvalue, however, represents a solution to $f(\lambda) = 1$ with $\lambda > \varepsilon_{max}$. Depending on the values of g_i^2, the outlying eigenvalue can be quite far removed from the others.

(c) With these choices for g_i and ε_i, the matrix **M** takes the special form

$$\mathbf{M} = \begin{pmatrix} \varepsilon + g^2 & g^2 & g^2 \\ g^2 & \ddots & g^2 \\ g^2 & g^2 & \varepsilon + g^2 \end{pmatrix}. \tag{4.30}$$

In this case, $\varepsilon_{min} = \varepsilon_{max} = \varepsilon$. We know from the graphical solution to part (b) of this problem that the $N - 1$ sandwiched eigenvalues are trapped between ε_{min} and ε_{max}. Therefore, we can infer that all of those $N-1$ eigenvalues now must be equal to ε. Equation 4.28 is not valid when $\lambda = \varepsilon$, but it can be used to find the remaining outlier eigenvalue, which we will call λ',

$$\sum_{i=1}^{N} \frac{g^2}{(\lambda' - \varepsilon)} = \frac{Ng^2}{(\lambda' - \varepsilon)} = 1, \tag{4.31}$$

or $\lambda' = \varepsilon + Ng^2$.

The sum of the diagonal elements of a matrix (i.e., its trace) provides a convenient check for these results. For reasons explained in linear algebra texts, the trace of the matrix $\mathbf{M}$ is equal to the sum of the eigenvalues. From Equation 4.30, the trace of $\mathbf{M}$ is $N(\varepsilon + g^2)$, which we can verify is indeed equal to the sum of the eigenvalues $(N - 1)\varepsilon + (\varepsilon + Ng^2)$.

(d) To determine the eigenvector $\vec{v}'$ associated with the outlying eigenvalue λ', we substitute $\lambda' = \varepsilon + Ng^2$ into Equation 4.26 to yield the simple expression

$$v_i' = \frac{C}{Ng}. \tag{4.32}$$

Each of the N elements of the vector $\vec{v}'$ has the same value, i.e., the index i does not appear on the right-hand side of Equation 4.32. To satisfy the normalization condition $\vec{v}' \cdot \vec{v}' = 1$, we rescale the eigenvector so that its elements are

$$v_i' = \frac{1}{\sqrt{N}}. \tag{4.33}$$

With this normalization, Equation 4.25 gives

$$C' = \sum_{i=1}^{N} g_i v_i' = g\sqrt{N}. \tag{4.34}$$

(e) If all of the g_i's are purely imaginary numbers, then each factor of g_i^2 that appears in Equation 4.28 is a real, negative number. In this case, Equation 4.28 becomes

$$-1 = \sum_{i=1}^{N} \frac{|g_i^2|}{(\lambda - \varepsilon_i)}. \tag{4.35}$$

We can construct an analogous graphical solution for Equation 4.35, but now we look for values of λ where the graph crosses -1, instead of $+1$. As before, $N - 1$ of the eigenvalues will be trapped between ε_{min} and ε_{max}. The outlier eigenvalue, however, will satisfy $\lambda < \varepsilon_{min}$, instead of $\lambda > \varepsilon_{max}$, as before.

Starting with Equation 4.23, it does not look very promising that a graphical solution will be available for this problem. Only after a few algebraic steps are performed, and Equation 4.28 is derived, does the opportunity arise.

It is interesting to compare the graphical solution of Example 4.3 with standard "brute force" numerical methods to solve for eigenvalues. As explained in detail in many linear algebra texts, eigenvalues are usually found by forming the determinant and solving the equation

$$\det(\mathbf{M} - \lambda\mathbf{I}) = 0. \tag{4.36}$$

For the matrix $\mathbf{M}$ given by Equation 4.20, this yields

$$\det\begin{pmatrix} 3-\lambda & \sqrt{12} & \sqrt{15} \\ \sqrt{12} & 4-\lambda & \sqrt{20} \\ \sqrt{15} & \sqrt{20} & 5-\lambda \end{pmatrix} = 0. \tag{4.37}$$

The resulting cubic equation in λ (called the characteristic equation for $\mathbf{M}$),

$$\lambda^3 - 18\lambda^2 + 57\lambda - 46 = 0, \tag{4.38}$$

can be solved either analytically or numerically. The results for the three roots are $\lambda_1 \approx 1.308$, $\lambda_2 \approx 2.473$, and $\lambda_3 \approx 14.22$, in agreement with the graphical solution. As the dimension N of the matrix increases, analytic solution for the eigenvalues becomes impossible, and numerical solution becomes progressively computationally burdensome. Consequently, the intuitive guidance provided by the graphical solution becomes increasingly valuable.

Finding the eigenvalues of matrices is a practical problem of considerable importance, with applications in diverse fields including physics and physical chemistry, economics, differential equations, and many others. Even so, the structure of the particular matrices in Equations 4.18 and 4.30 appears to be contrived, so that finding their eigenvalues might not seem very useful. The eigenvalues for this class of matrix, however, have several important applications. For example, in quantum mechanics, the eigenvalues of the energy matrix (also called the Hamiltonian) correspond to discrete energy levels of a system. A quantum mechanical calculation for the energy levels of the nucleus of an atom yields this type of matrix. The attractive nuclear force leads to a matrix of the form in Example 4.3e and provides a reason why one energy level can lie far below the others, which can be tightly bunched. The book by Brown (1967) describes the nuclear "schematic model," which uses this concept and provides great insight into the structure and behavior of atomic nuclei.

This type of matrix also helps to explain the "pairing" phenomenon that is associated with superconductivity observed in some metals at low temperatures. Two electrons normally repel each other due to electrostatic repulsion, but in a superconductor, they can attract to form a Cooper pair. The two electrons in the Cooper pair are correlated; their spin angular momentum is oriented in opposite directions, and they travel together, although with a relatively large separation (up to hundreds of nanometers). The details are beyond the scope of this book, but the energy or Hamiltonian matrix that governs the quantum mechanical problem can be modeled by a matrix with attractive (i.e., negative) off-diagonal elements, similar to the matrix in Example 4.3. This yields a quantum state with particularly low energy and is a basis for the BCS theory of superconductivity named after the American physicists John Bardeen,

Leon Cooper, and Robert Schrieffer, who were awarded the Nobel Prize in 1972 for their work.

4.4 SYMMETRY TO SIMPLIFY EQUATIONS

In everyday language, "symmetrical" often means well balanced, particularly in regard to bilateral symmetry. In mathematics and science, the word is used in a more specific sense. As described in Weyl (1983; a much more detailed discussion of symmetry is offered in this book), we say something is "symmetrical" if it appears unchanged after being subjected to some type of operation. Consistent with everyday usage, something with bilateral symmetry appears unchanged after reflection in a mirror. There are many other types of symmetry operations, however. For example, if a wheel looks the same after it is rotated through a certain number of degrees about its axis, then it has a rotational symmetry.

Symmetry is a very broad field, and group theory is an entire branch of mathematics devoted to its study. Also, symmetry of a physical system described by quantum mechanics has the interesting property that it is associated with a corresponding conserved quantity. For example, the rotational symmetry mentioned earlier is related to the conservation of angular momentum. These and other aspects of symmetry are beyond the scope of this book. Instead, we will provide a few examples of how exploiting symmetry can simplify equations, which is a powerful method in its own right. The first example is taken from electrical engineering.

EXAMPLE 4.4

Consider a set of 12 resistors, each with resistance $R = 1\Omega$. One resistor is placed along each of the 12 edges of a cube to form a network. Electrical current enters at one corner a of the cube and exits at the diagonally opposed corner h, as shown in Fig. 4.8. (To aid visualization, we can lay the resistors onto a flat surface, as in Fig. 4.9.) Use symmetry considerations to find the equivalent resistance R_{eq} between points a and h of this network of 12 resistors. Compare the amount of effort required to solve this problem using symmetry considerations, as opposed to direct calculation.

ANSWER

The same amount of electrical current that enters the cube must exit the cube. From this, we can infer that

$$I_{in} = I_{out} = I. \tag{4.39}$$

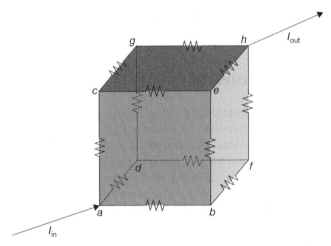

FIG. 4.8. Twelve 1Ω resistors are placed along the edges of a cube to form a network. Symmetry considerations can simplify the task of finding the equivalent resistance between opposite corners a and h of the cube.

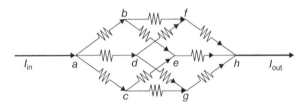

FIG. 4.9. For easier visualization, the network of resistors in Fig. 4.8 can be laid flat.

We can define 12 values of the current I_{ab}, I_{ac}, ..., I_{fh}, I_{gh}, i.e., one for each edge of the cube. The voltage drop across any resistor is equal to the product of the current passing through it and its resistance, for example, on the *ab*-edge of the cube:

$$V_{ab} = I_{ab}R. \qquad (4.40)$$

Kirchhoff's rules tell us that (1) the sum of the currents flowing into any junction (such as one of the eight corners of the cube) must equal the sum of currents leaving that junction and (2) the sum of the voltage drops around any closed loop in a circuit is equal to zero. For the second rule, the sign of the voltage drop is determined by the direction of the current.

Applying Kirchhoff's first rule to each corner of the cube yields eight equations. We can generate a sufficient number of equations to solve for the individual currents by applying Kirchhoff's second rule to current loops, such as those on the six faces of the cube. The resultant system of equations, however, is messy to solve.

This problem can be greatly simplified by exploiting its inherent symmetry. When the current I exits corner a, the three available paths (ab, ac, or ad on Fig. 4.8) are equivalent because all of the resistors have the same value, and the cube has symmetry under a 120° rotation about the ah-diagonal. Therefore, symmetry requires that the current splits equally among these three paths. Applying Kirchhoff's first rule and Equation 4.39 to the junction a yields

$$I_{ab} = I_{ac} = I_{ad} = \frac{I_{in}}{3} = \frac{I}{3}. \qquad (4.41)$$

Similarly, the three currents converging onto point h satisfy the relationships

$$I_{eh} = I_{fh} = I_{gh} = \frac{I_{out}}{3} = \frac{I}{3}. \qquad (4.42)$$

How does the current divide after it enters point b, c, or d? We can see from Fig. 4.9 that at each of these junctions, the current has two available paths. For example, at point b, the current I_{ab} divides into I_{be} and I_{bf}. As is apparent from Fig. 4.8, symmetry considerations require that the current divides evenly. Therefore, Kirchhoff's first rule and Equation 4.41 applied to point b yields

$$I_{be} = I_{bf} = \frac{I_{ab}}{2} = \frac{I}{6}. \qquad (4.43)$$

The total voltage drop V_{tot} of the network can be calculated along any path that goes from point a to point h. There are many such paths, the simplest of which covers three edges. For example,

$$V_{tot} = I_{ab} R + I_{bf} R + I_{fh} R. \qquad (4.44)$$

Using Equations 4.41–4.44, we find that

$$V_{tot} = \left(\frac{1}{3} + \frac{1}{6} + \frac{1}{3}\right) IR = \frac{5}{6} IR. \qquad (4.45)$$

The equivalent resistance R_{eq} of the network satisfies the relation

$$V_{tot} = IR_{eq}, \qquad (4.46)$$

so the desired result is $R_{eq} = 5/6\ R = 0.833\,\Omega$. We can check this result by returning to Equations 4.41–4.43 and then by verifying that Kirchhoff's first rule is satisfied at any junction, and Kirchhoff's second rule is satisfied for any current loop.

The problem in Example 4.4 has a high degree of symmetry because of the geometry of the cube and the fact that all the resistors have the same value $R = 1\Omega$. Equation 4.41 holds because there is no reason for the value of one of the three currents to differ from the other two. For example, with the cube depicted in Fig. 4.8, if we swap the labels of corners b and c, the network would still operate the same, i.e., swapping of the labels is a symmetry operation due to the rotational symmetry about the ah-axis mentioned earlier. The relabeling of the corners does not change the numerical values of the electrical currents I_{ab} or I_{ac}. If, however, one of the resistors had a different value from the others, say $R_{ab} = 2\Omega$, the symmetry would be broken and the simplifying arguments would no longer be valid.

Examples 4.2 and 4.3 explored graphical solutions extracted from plots of functions. Example 4.4 dealt with symmetry. We conclude this chapter by combining some of these ideas to show that useful information about a function can be extracted when its graph displays symmetry.

A well-known example is reflection symmetry of the graph of a function about the y-axis. A function is called *even* if it is unchanged by reflection about the line $x = 0$ (the y-axis), like x^2 or $\cos x$. We write

$$f_e(-x) = f_e(x). \qquad (4.47)$$

The graph of an *odd* function like x^3 or $\sin x$ is antisymmetric when reflected:

$$f_o(-x) = -f_o(-x). \qquad (4.48)$$

Note that if an odd function is defined at the origin, its value must be zero there: $f_o(0) = 0$.

Calculation of a definite integral over the symmetric interval $[-L, L]$ often can be simplified by decomposing the definite integral into two parts,

$$\int_{-L}^{L} f(x)\,dx = \int_{-L}^{0} f(x)\,dx + \int_{0}^{L} f(x)\,dx, \qquad (4.49)$$

and then exploiting the reflection symmetry of the specific function. With an even function, the two contributions to the area represented by the integral are equal, so

$$\int_{-L}^{L} f_e(x)\,dx = 2\int_{0}^{L} f_e(x)\,dx. \qquad (4.50)$$

For an odd function, the two contributions to the area represented by the integral in Equation 4.49 exactly cancel, so the integral is zero:

$$\int_{-L}^{L} f_o(x)\,dx = 0. \qquad (4.51)$$

It is easy to show that the *multiplication* of even and odd functions follows rules analogous to the *addition* of even and odd integers. In particular, if the product of an even and an odd function is $g_o(x)$, then that function is odd:

$$f_e(x) \times f_o(x) = g_o(x). \tag{4.52}$$

Analogously, the sum of an even and an odd integer is an odd integer.

These rules can be used to simplify the calculation of trigonometric series. Almost any periodic (i.e., repeating) function can be decomposed into an infinite series of sines, cosines, and a constant:

$$f(x) = \frac{a_0}{2} + \sum_{n=1}^{\infty} a_n \cos\frac{n\pi x}{L} + \sum_{n=1}^{\infty} b_n \sin\frac{n\pi x}{L}. \tag{4.53}$$

This expansion is called a Fourier series after Jean Baptiste Joseph Fourier, a French mathematician. Figure 4.10 shows several examples of periodic functions and their Fourier series.

The spatial period of the function $f(x)$ is $T = 2 \times L$, that is, $f(x) = f(x \pm T) = f(x \pm 2T)$, and so on. The a_n's and b_n's are called the Fourier coefficients of the series, which can be obtained by evaluating definite integrals over one period:

$$a_n = \frac{1}{L} \int_{x=-L}^{L} f(x)\cos\frac{\pi n x}{L} dx \tag{4.54}$$

and

$$b_n = \frac{1}{L} \int_{x=-L}^{L} f(x)\sin\frac{\pi n x}{L} dx \tag{4.55}$$

Depending on the specific function $f(x)$, the definite integrals in Equations 4.54 and 4.55 can be messy to evaluate, so the use of symmetry to simplify the problem is helpful. Because cosine is an even function and sine is odd, symmetry properties can greatly simplify calculation of certain Fourier coefficients. According to Equations 4.51 and 4.52, all of the coefficients of the sine terms are zero in the Fourier series of an even function

$$b_n = 0, \quad n = 1, 2, 3, \dots \text{(even function)}, \tag{4.56}$$

because the integration of the resultant odd function over a symmetric interval is zero.

Similarly, for an odd function, all of the a_n's are zero.

$$a_n = 0, \quad n = 0, 1, 2, 3, \dots \text{(odd function)}. \tag{4.57}$$

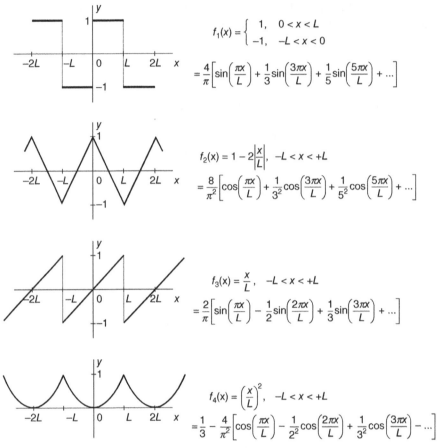

FIG. 4.10. Four periodic functions and their Fourier series. The period of each function is $2L$.

Equations 4.56 and 4.57 should not be very surprising. An even periodic function is therefore composed exclusively of a series of even terms (cosines and a constant) and is called a Fourier cosine series. An odd periodic function is composed exclusively of a series of odd terms, which is called a Fourier sine series. If a function is neither even nor odd, both types of terms in Equation 4.53 are present.

The following problem illustrates how we can infer still more about which Fourier coefficients are zero by examining other symmetry present in $f(x)$.

EXAMPLE 4.5

(a) Based on their graphs, determine whether Equation 4.56 or 4.57 applies to any of the functions plotted in Figure 4.10.

(b) How can you tell simply by looking at the graph whether the constant term in the Fourier series a_0 is zero?

(c) Note that the even harmonics (i.e., $n = 2, 4, 6 \dots$) are missing in the Fourier series of the bipolar square wave $f_1(x)$. Develop rules based on the symmetry of the graph to determine when the even or odd harmonics will be missing.

ANSWER

(a) The functions $f_1(x)$ and $f_3(x)$ are odd and consistent with Equation 4.56; their Fourier series are missing the constant term and all of the cosine terms. Similarly, the functions $f_2(x)$ and $f_4(x)$ are even, so their Fourier series are missing all the sine terms.

(b) According to Equation 4.54,

$$a_0 = \frac{1}{L} \int_{x=-L}^{L} f(x)\,dx, \tag{4.58}$$

which is twice the average of $f(x)$ over one period $T = 2L$. This same result can also be seen from Equation 4.53, because the averages over one period of all the sines and cosines are zero because of their oscillation about the x-axis. From the graphs in Fig. 4.10, $f_4(x)$ is the only one of the four functions that has a nonzero average over $-L < x < L$. It is also the only function that has a nonzero constant term a_0 in its Fourier series.

As can be seen from Equation 4.57, every odd function must have a zero average over one period. As illustrated by $f_2(x)$ and $f_4(x)$, an even function might or might not have a zero average over one period. The same holds true for functions that are neither even nor odd.

(c) A convenient way to determine when certain coefficients will be zero in a Fourier cosine or sine series is to examine the reflection symmetry of the graph about the line $x = L/2$ in the interval $0 < x < L$. For example, consider the bipolar square wave $f_1(x)$, which is an odd function so it has a Fourier sine series. Using Equation 4.55, the coefficient for the harmonic b_2 is given by

$$b_2 = \frac{1}{L} \int_{x=-L}^{L} f_1(x)\sin\frac{2\pi x}{L}\,dx. \tag{4.59}$$

Equation 4.59 can be evaluated directly and gives zero, but the same result can be obtained more simply using the symmetry apparent from Fig. 4.11. Because $f_1(x)$ is an odd function *and* is symmetric about the line $x = L/2$ over the interval $0 < x < L$, each positive contribution to the integral from $0 \le x \le L/2$ for b_2 is exactly canceled by an equal, negative contribution

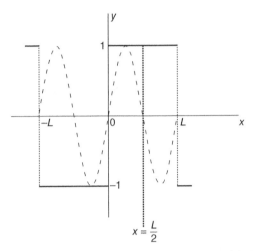

FIG. 4.11. Pictorial argument based on the symmetry of $f_1(x)$ (solid line) about $x = L/2$ that the coefficients of the even harmonic terms in its Fourier series are zero. The dashed line is a plot of $y = \sin(2\pi x/L)$.

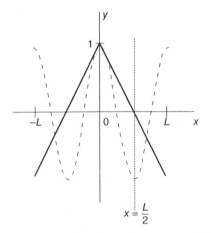

FIG. 4.12. Pictorial argument based on the antisymmetry of $f_2(x)$ (solid line) about $x = L/2$ that the coefficients of the even harmonic terms in its Fourier series are zero. The dashed line is a plot of $y = \cos(2\pi x/L)$.

from $L/2 \leq x \leq L$, so the net result is zero. Similar symmetry arguments show that all the remaining even harmonics of $f_1(x)$ are zero as well, i.e., $b_4 = b_6 = b_8 = \ldots = 0$.

The sawtooth wave function $f_2(x)$ is an even function whose Fourier cosine series is missing the even harmonics, i.e., $a_4 = a_6 = a_8 = \ldots = 0$. From Equation 4.54, we can calculate Equation 4.60 directly:

TABLE 4.1. Reflection Symmetry of the Graph of an Even or Odd Periodic Function Can Determine That Some of the Coefficients in Its Fourier Series Are Zero, Avoiding a Great Deal of Computational Effort.

	Even	Even, Symmetric about $L/2$	Even, Antisymmetric about $L/2$	Odd	Odd, Symmetric about $L/2$	Odd, Antisymmetric about $L/2$
a_n				0	0	0
$a_1, a_3, a_5\ldots$		0		0	0	0
$a_2, a_4, a_6\ldots$			0	0	0	0
b_n	0	0	0			
$b_1, b_3, b_5\ldots$	0	0	0			0
$b_2, b_4, b_6\ldots$	0	0	0		0	

$$a_2 = \frac{1}{L} \int_{x=-L}^{L} f_2(x) \cos\frac{2\pi x}{L} dx. \tag{4.60}$$

Again, symmetry considerations simplify the work. Figure 4.12 illustrates pictorially why $a_2 = 0$. While the sawtooth wave $f_2(x)$ is an even function when considering reflections about $x = 0$, it is nonetheless antisymmetric when considering reflections about the line $x = L/2$. The function $\cos(2\pi x/L)$ is symmetric about $x = L/2$. Consequently, the net result of the integration of the product of the two functions over the interval $0 \le x \le L/2$ is zero.

These methods can be extended to show other related results, as summarized in Table 4.1. Exercises 4.5 and 4.8 explore the further use of symmetry to simplify the calculation of Fourier coefficients.

Example 4.5 illustrates the value of using a graph in order to visually determine those symmetries that can simplify the problem of calculating certain Fourier coefficients. The ear can be sensitive to symmetry in the waveform of a periodic function as well. The sense of hearing detects the temporal variation in air pressure (relative to atmospheric pressure), which can be described by a waveform $P(t)$, with the sound intensity proportional to $P^2(t)$. Assuming that $P(t)$ is zero at time $t = 0$ and repeats with period T, we can decompose $P(t)$ into a Fourier sine series:

$$P(t) = \sum_{n=1}^{\infty} b_n \sin\left(\frac{2\pi nt}{T}\right) = \sum_{n=1}^{\infty} b_n \sin(2\pi f_n t). \tag{4.61}$$

We have replaced the spatial variable x in Equation 4.53 with the temporal variable t. We also have replaced the spatial period $2L$ by the temporal period

T (measured in units of seconds). The $n = 1$ term is called the *fundamental* and has the lowest frequency $f_1 = 1/T$ (measured in hertz). The higher-order terms are the harmonics with higher frequencies that are an integer multiple of f_1:

$$f_n = \frac{n}{T}. \tag{4.62}$$

The specific shape of the waveform $P(t)$ permits our ears to distinguish between various instruments playing notes of the same pitch, i.e., with the same fundamental frequency f_1. For example, our ears can distinguish between a clarinet and a flute, both playing the note A with $f_1 = 440\,\text{Hz}$. While the pitch is determined by f_1, the relative strengths of the various Fourier coefficients b_n for $n \geq 2$ determine the unique tonal characteristics of each instrument.

Generally, the amplitude of the coefficients b_n decreases for larger values of n. In some types of musical instruments, the coefficients for odd n might be very small, or zero. The specific amplitudes depend on the particular musical instrument and, in some cases, can be determined by examining its structure for symmetry. Expectations about the general properties of the coefficients can be found by applying some basic principles from acoustic physics. We will introduce a few basic principles here, but a much more thorough discussion of acoustic physics is contained in Fletcher (2005).

A column of air in a pipe (i.e., a hollow cylinder) can resonate, producing a musical tone. An end of the pipe can be open or closed. At an open end, boundary conditions require that the $P(t)$ (the variation of pressure from the ambient atmospheric air pressure) of the resonating air column be zero. This is called a "pressure node." At a closed end, the amplitude of $P(t)$ has a maximum, or a "pressure antinode." As illustrated in Fig. 4.13, for a cylinder that is open at one end and closed at the other, these boundary conditions require that only an odd number of quarter wavelengths fit into the length ℓ of the cylinder, or

$$(2k-1)\frac{\lambda_n}{4} = \ell, \quad n = (2k-1), \quad k = 1, 2, 3 \ldots, \tag{4.63}$$

where λ_n is the allowed wavelength. Because the product of frequency and wavelength is equal to the speed of sound v_{sound}, Equation 4.63 can be rearranged to provide the allowed frequencies:

$$f_n = \frac{v_{\text{sound}}}{\lambda_n} = \frac{(2k-1)v_{\text{sound}}}{4\ell}, \quad k = 1, 2, 3 \ldots. \tag{4.64}$$

Comparing Equations 4.62 and 4.64, the fundamental frequency is

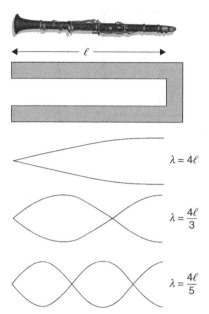

FIG. 4.13. A cylinder that is open at one end can serve as a simplified model of a clarinet. Standing sound waves produce a pressure maximum at the closed end and a zero (node) at the open end. These boundary conditions require that an odd number of quarter wavelengths fit along the length of the pipe.

$$f_1 = \frac{v_{\text{sound}}}{4\ell} \quad \text{(pipe closed at one end)}, \qquad (4.65)$$

and the even harmonics corresponding to $f_2, f_4, f_6, \ldots$ are missing.

This physical configuration can be used to model a clarinet, with the mouthpiece considered a closed end. Experimental measurements show that the even Fourier coefficients b_2, b_4, etc., are usually quite small for a waveform produced with a clarinet, so the model is reasonably good. When a hole on the side of the clarinet is exposed by pressing a key, the length ℓ from the closed end to the "open" end is changed, which changes pitch according to Equation 4.64.

When a column of air resonates in a pipe that is open at both ends, the boundary conditions are very different from those described in connection with Equation 4.63. The cylinder open at both ends has air pressure nodes at both ends. Using diagrams analogous to Fig. 4.13, it can be shown that the length ℓ of the open pipe can contain an integer number of half wavelengths, so that

$$f_1 = \frac{v_{\text{sound}}}{2\ell} \quad \text{(pipe open at both ends)}. \qquad (4.66)$$

Therefore, both even and odd harmonics are allowed with this boundary condition. A flute is a woodwind instrument that is well approximated by a

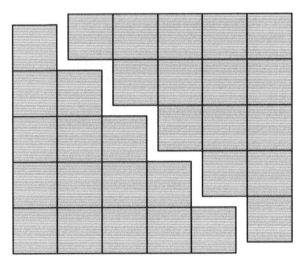

FIG. 4.14. Figure for exercise 4.1.

pipe open at both ends. Comparing Equations 4.65 and 4.66, we expect the pitch of the clarinet to be approximately one octave lower than for a flute of the same length. Experimental measurements support this difference in pitch, i.e., for instruments of similar length, the fundamental frequency for a clarinet is approximately half that of a flute. Also, the even Fourier coefficients b_2, b_4, ... are larger for the musical tones produced by a flute.

The lack of even harmonics in the clarinet's waveform helps the human ear–brain system to distinguish its sound from that of a flute when the same fundamental frequency f_1 is played. This is one way in which the ear is sensitive to symmetry.

EXERCISES

(4.1) Consider the following sequence of numbers

$$T_1 = 1,$$
$$T_2 = 1 + 2 = 3,$$
$$T_3 = 1 + 2 + 3 = 6,$$
$$T_4 = 1 + 2 + 3 + 4 = 10,$$
$$T_k = 1 + 2 + 3 + \ldots + k.$$

Show how the diagram in Fig. 4.14 demonstrates the identity $T_k = \frac{1}{2}k(k+1)$. Explain why T_k is called a sequence of "triangular" numbers.

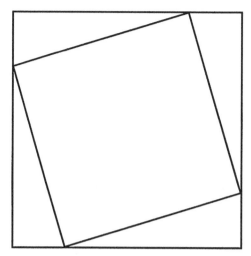

FIG. 4.15. Figure for exercise 4.2.

(4.2) Show how Fig. 4.15 is related to the Pythagorean theorem $a^2 + b^2 = c^2$. To what extent do you think that this diagram represents a "proof" of the theorem? A similar diagram is found in the mathematical text *Zhou Bi Suan Jing*, dating from the Zhou Dynasty in China (1046–256 BC).

(4.3)

(a) Assuming that Equation 4.14 holds, show that from Equation 4.16 if h happens to be an identity function, then g is simply the inverse of f, and vice versa. Interpret this special case geometrically on a Jones plot by substituting an identity line for the function h.

(b) Relax the assumption that Equation 4.14 holds, so that the output from the printer can differ from the original. Consider the graphs in Fig. 4.16. Which one corresponds to a printout that is lighter than the original?

(c) A darker printout than the original?

(d) A printout with higher contrast than the original?

(e) Draw a graph for A_{out} versus A_{in} that will produce a lower-contrast print.

(4.4) An odd function as defined by Equation 4.48 is antisymmetric under reflection in the y-axis. Show geometrically that an odd function is symmetric with respect to *rotation* in the xy-plane by $180°$ about the origin.

(4.5) Consult a table of Fourier series that has accompanying plots of the functions and verify more cases of the symmetry properties listed in Table 4.1. *Schaum's Mathematical Handbook of Formulas and Tables* by Murray R. Spiegel, ISBN 0070382034, among others, contains an extensive table.

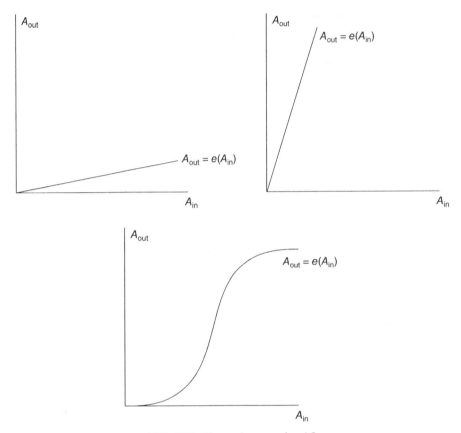

FIG. 4.16. Figure for exercise 4.3.

(4.6) Equation 4.4 also has use in the theory of equations of variables that take on integer values. Show that the value of N provided in Equation 4.4 is equal to the number of distinct solutions to the equation

$$x_1 + x_2 + \ldots + x_r = n, \tag{4.67}$$

where the values of the x's in Equation 4.67 are nonnegative integers.

(4.7) Without graphing them, test each equation in Equation 4.68 for symmetry about:
(a) the x-axis (replace y with $-y$ and determine if the equation is unchanged);
(b) the y-axis (replace x with $-x$ and determine if the equation is unchanged);
(c) the line $x = y$ (swap x and y and determine if the equation is unchanged).

$$x^2 + y^2 = 1,$$

$$xy = 4,$$

$$y = 3x^2 + 2x + 2,$$

$$y = x^4 + \cos(2x).$$

(4.68)

(4.8) Consider the periodic function

$$f(x) = \begin{cases} 1 - \dfrac{12|x|}{L}, & |x| \le \dfrac{L}{6} \\ \dfrac{12|x|}{L} - 3, & \dfrac{L}{6} < |x| \le \dfrac{L}{3} \end{cases}.$$

(a) Sketch a plot of $f(x)$. Show that the graph of f has a sawtooth shape. Although $f(x)$ is periodic with $T = 2L$, show that the shortest period of f is $L/3$.

(b) Use symmetry to show that all the Fourier coefficients b_n are equal to zero, and a_0 is zero.

(c) Use symmetry to show that the third Fourier cosine coefficient a_3 is zero.

(d) Sketch a graph for a function for which all the b_n's are equal to zero and the kth Fourier cosine coefficient is also zero.

(4.9) Referring to Table 4.1, consider an even function that is symmetric about the line $x = L/2$. Is $2L$ the shortest period for this function? If not, find a shorter period. Can you draw an analogous conclusion about an odd function that is antisymmetric about the line $x = L/2$?

(4.10) Consider a suitcase with a four-digit combination lock. Each of the four digits ranges from 0 to 9, so there are a total of 10,000 possible distinct numerical strings ranging from 0-0-0-0 to 9-9-9-9.

(a) Using symmetry arguments, show that considering all of the possible numerical strings, the digit "7" occurs a total of 4000 times. (The string 7-8-7-7 counts as three occurrences of the digit 7.) Hint: there are 10,000 distinct strings, with four digits per string, yielding a total of 40,000 distinct digits. Symmetry suggests that none of the 10 digits $0, 1, 2, \dots, 9$ appear more frequently than any other, which leads to the desired result.

(b) Generalize the result in part (a) to show that the number of times the digit "7" occurs in all possible numerical strings generated with an n-digit lock is given by $n \times 10^{n-1}$. Test this general result for the special cases $n = 1$ and $n = 2$ by explicitly counting the occurrences of the digit 7 for the numbers 0–9 and 0–99, respectively.

(4.11) Consider the well-known algebraic expression $(a + b)^2 = a^2 + b^2 + 2ab$.
(a) Draw a diagram of a square with side-length $a + b$ and interpret the areas represented in the diagram based on the algebraic expression.
(b) Draw a diagram of two concentric circles with radii R and $R + \Delta r$. Interpret the areas represented in this diagram based on the same algebraic expression. Hint: observe that

$$\pi(R + \Delta r)^2 = \pi R^2 + 2\pi \Delta r \left(R + \frac{\Delta r}{2} \right).$$

REFERENCES

Brown GE. 1964. *Unified Theory of Nuclear Models and Forces*. New York: Interscience Publishers.

Dehaene S, Izard V, Pica P, and Spelke E. 2006. Core knowledge of geometry in an Amazonian indigene group. Science 311: 3181–3184.

Fletcher NH. 2005. *The Physics of Musical Instruments*. New York: Springer.

Weyl H. 1983. *Symmetry*. Princeton, NJ: Princeton University Press.

FURTHER READING

Dunham W. 1990. *A Journey through Genius: The Great Theorems of Mathematics*. New York: John Wiley & Sons.

Those who are interested in the historical development of mathematics, such as the work of Archimedes to determine the area of a circle, will find the books by William Dunham of great interest. In particular, we recommend this book.

Feynman RP, Leighton RB, and Sands M. 2005. *The Feynman Lectures on Physics*. Boston: Addison-Wesley.

Perhaps no modern scientist is more closely associated with the use of diagrams to represent and to help solve complicated equations than Richard P. Feynman, an American physicist. The classic three volume set presents a course on physics with a wealth of physical and geometric intuition.

Brown JR. 1999. *Philosophy of Mathematics: An Introduction to a World of Proofs and Pictures*. London: Routledge.

The relationship between pictures and mathematical proofs is discussed in this book.

5

ESTIMATION AND APPROXIMATION

Most problems in science and engineering can be modeled with equations, but only a few yield equations for which we can find an exact, analytic solution. In this chapter, we discuss estimation and approximation techniques that can be applied to many of the remaining problems. Even if the exact solution is attainable, an approximate solution might suffice and is usually much easier to find. The array of estimation and approximation approaches is very broad, and here we limit ourselves to a few selected methods to give a flavor for this topic.

One use of the word "estimation" is to describe the process with which we make quick, mental calculations. Section 5.1 contains an example illustrating how using powers of two provides simple and accurate numerical estimates for geometric sequences (also known as geometric progressions). Another use of the word "estimation" is to describe the technique of making order-of-magnitude guesses for numerical values based on simple principles and commonsense reasoning. These techniques are illustrated with example problems in Sections 5.2 and 5.3. The study of such problems has been advocated to help remedy "innumeracy" (i.e., numerical illiteracy) or a lack of "number sense" manifested by difficulty in determining whether the order of magnitude of a numerical quantity is plausible.

The remaining three sections of this chapter deal with approximation methods. Several analytic methods that can be used to approximate the value of definite integrals are discussed in Section 5.4. Basic aspects of perturbation

Thinking About Equations: A Practical Guide for Developing Mathematical Intuition in the Physical Sciences and Engineering, by Matt A. Bernstein and William A. Friedman
Copyright © 2009 John Wiley & Sons, Inc.

analysis are introduced in Section 5.5. The chapter concludes with a qualitative discussion of the importance of identifying and isolating the most important variables in a problem. These ideas are illustrated by discussions of Brownian motion and the nuclear optical model.

5.1   POWERS OF TWO FOR ESTIMATION

We will begin with a simple technique that can help improve the accuracy of certain estimates and can be quickly carried out without the aid of a calculator. Many processes such as population growth and investment income from compound interest follow a geometric sequence, i.e., their values increase exponentially. When estimating numerical results for these sequences, we have a choice of using powers of 2 (i.e., 2, 4, 8, 16, ...), powers of 10 (i.e., 10, 100, 1000, ...), or powers of any other base. When analyzing problems using calculus, we usually choose powers of Euler's constant e, because only $f(x) = e^x$ has the property that $f'(x) = f(x)$. However, for quick, everyday numerical estimates, powers of 2 or 10 are usually the most convenient. The next example illustrates that using powers of two is often preferred because the smaller step size can improve accuracy.

EXAMPLE 5.1

Consider a tax-free bond yielding a fixed annual percentage rate r, with the interest compounded annually.

(a) Show that the number of years n_2 that it takes for the principal P_0 to double is approximated by

$$n_2 \approx \frac{72}{r}. \qquad (5.1)$$

Equation 5.1 is sometimes called the *rule of 72*. For example, if the annual percentage rate is 6%, then the value of the bond will double in approximately 72/6 = 12 years.

(b) Show that we alternatively could have developed a "rule of 240" for the number of years that it would take the bond to increase in value by a factor of 10. Discuss why, in practice, the rule of 72 is used instead.

ANSWER

(a) The value of the bond $P(n)$ after n years is given by

$$P(n) = P_0(1+r)^n. \tag{5.2}$$

The value of the bond doubles when $P(n_2)/P_0 = 2$. Taking the logarithm of each side of Equation 5.2 and solving for n yields

$$n_2 = \frac{\ln 2}{\ln(1+r)}. \tag{5.3}$$

For $|r| < 1$, the denominator can be expanded: $\ln(1 + r) = r - r^2/2 + \dots$. Neglecting cubic and higher order terms in r, which is appropriate when $|r| \ll 1$,

$$n_2 \approx \frac{\ln 2}{r(1-r/2)}. \tag{5.4}$$

Equation 5.4 is a very good estimate, but it is still somewhat difficult to evaluate without a calculator. The numerical values, however, work out nicely if we assume that $r \approx 0.075$ or 7.5%. This value is in the range of interest rates typically used in financial transactions. Then the term $(1 - r/2) = 0.9625$, and

$$n_2 \approx \frac{\ln 2}{r(1-r/2)} = \frac{\ln 2}{r \times 0.9625} \approx \frac{0.72}{r}. \tag{5.5}$$

The rule of 72 can be applied to any process with exponential growth, but its accuracy decreases as the growth rate deviates from 7.5%. For an interest rate of 6%, the rule of 72 estimates that the bond value doubles in 12 years, which is close to the value of 11.896 years obtained from Equation 5.3.

(b) We can go through similar steps and show that the number of years required for an increase by a factor of 10 is given by

$$n_{10} \approx \frac{\ln 10}{r(1-r/2)} = \frac{\ln 10}{r \times 0.9625} \approx \frac{2.39}{r}. \tag{5.6}$$

To make the mental arithmetic easier, we will round up the numerator of Equation 5.6 to 2.4. So, at 6% annual rate, our bond value will increase by approximately a factor of 10 in $240/6 = 40$ years. The precise value in this case is 39.517 years, so the estimate is approximately half a year too long. The rule of 72 is widely used, but "the rule of 240" has never caught on, and for several good reasons. First, investors rarely hold a bond long enough for its value to increase by a factor of 10. Second, which is the main point of this example, estimating with powers of two is usually more accurate. Clearly, the value of a variable undergoing geometric growth doubles more frequently than it increases by a factor of 10. Therefore, using powers of two typically has the advantage of being closer to one of those milestones. This idea is further explored in exercise 5.2. Whenever conversion between powers of 2 and powers of 10 is required, as in exercise 5.7, it is useful to remember the handy fact that $2^{10} = 1024 \approx 10^3$.

Perhaps, computer science, more than any other field, makes extensive use of powers of two. This is not surprising, because digital computers perform calculations with binary arithmetic. The digits 0 and 1 provide a natural representation for the off/on states of the electronic hardware components. Powers of two also seem to appear in other areas of computer science. For example, Moore's law, named after the American industrialist Gordon E. Moore, describes the historic trend that the number of transistors that can be placed on a computer chip (i.e., an integrated circuit) has doubled every 2 years. Interestingly, this "law" has proved quite accurate for the first half century of the integrated circuit, possibly because over time, an expectation has developed for achieving this growth.

When using powers of two in the computer science arena, it should be noted that the common prefixes kilo, mega, giga, tera, etc., have different numerical values when describing data storage than in other fields of science and engineering. One megawatt is *exactly* equal to 1,000,000 W, while 1 megabyte of storage is only *approximately* equal to 1,000,000 B. By definition, it is $2^{20} = 1,048,576$ B, which is 5% more than one might guess! For 1 terabyte (2^{40} B), confusing the two usages introduces a substantial error, because a terabyte is 10% larger than the 10^{12} B suggested by its name.

5.2 FERMI QUESTIONS

Next, we examine methods for making rough estimates based on reasonable assumptions. An answer to such a problem is sometimes called a ballpark estimate, a "guesstimate," a plausible estimate, or a back-of-the-envelope calculation. This type of estimation can be applied to analyze many day-to-day situations. A well-known problem credited to the twentieth-century Italian and American physicist Enrico Fermi is to estimate the number of piano tuners who were employed in the city of Chicago. As described in a letter to the editor by Philip Morrison in the August 1963 issue of the *American Journal of Physics*, Fermi often posed and answered these types of questions in order to develop estimation skills. As a result, they have become known as "Fermi problems" or "Fermi questions."

EXAMPLE 5.2

Rochester, Minnesota is a small city with a population of approximately 100,000 people. Estimate the total number N_{est} of hair salons and barbershops that are located within the city.

ANSWER

The estimate can be developed by decomposing the answer into a series of factors and then by making independent estimates for each one of them. The estimates for some of the factors might be too high, and others too low. Unless we are unlucky (or perhaps biased, in the statistical sense), some of these individual errors cancel, reducing the overall error of the entire estimate.

Given the population P, we will assume that a segment of the population $f \times P$ visits a salon or a barbershop on a regular basis. The fraction $f < 1$ accounts for infants and others who do not use the service. If, for a given individual, the average time interval between visits is D days, then we can calculate that a total of $f \times P/D$ customers visit a salon or a barbershop on an average day.

Suppose further, on average, each shop works on C haircuts concurrently, is open o hours per day, and that the duration of the average appointment is H hours. Balancing the demand for customer visits with the available capacity, we arrive at

$$\frac{f \times P}{D} = \frac{N_{est} \times C \times o}{H}. \tag{5.7}$$

Notice that both sides of Equation 5.7 have units of days^{-1}. Next, we take some educated guesses about the various factors in Equation 5.7. We will suppose that on average, the shop works on $C = 3$ haircuts at any given time and is open $o = 8\,\text{h}$ per day, and that an average appointment duration is $H = 0.5\,\text{h}$. We will also estimate the average interval between haircuts to be 35 days and use $f = 0.8 = 80\%$ for the fraction of the population that uses the service. From Equation 5.7, we arrive at the estimate $N_{est} = 47$ salons. Looking through a Rochester phone directory for the year 2007, we find the actual count of $N_{true} = 67$. So the rough estimate is accurate to approximately 30%, which is not too bad for this type of calculation.

Numerical values and statistics are frequently reported in the news media and by other sources. Posing and solving Fermi questions help us to develop skills that can quickly filter these data for plausibility. Misuse of numerical data can have serious consequences, but an amusing example is cited in De

Robertis (2000; Michael De Robertis has advocated the value of teaching Fermi questions in the classroom, starting at an early age, and this compilation is available online as an e-book). An advocate for the benefits of eating peanut butter once claimed that each American eats, on average, 100,000 peanut butter sandwiches over the course of a lifetime. Assuming an average life expectancy of 72 years, this works out to nearly four peanut butter sandwiches per person per day, a number that is too high to be plausible. Such misuse of numerical values not only desensitizes the listener but also eventually undermines the credibility of the speaker.

There is no single correct approach or answer for a Fermi question like Example 5.2. Notice that the methodology illustrated by Example 5.2 is very different from the science of sampling used in polling. A pollster might be asked to predict who will win an upcoming election. Because it is not practical to interview every eligible voter, the pollster interviews only a subset of them. Using the tools of probability and statistics, a quantitative margin of error for the polling results can be established based on the sample size relative to the number of voters.

The methods of Example 5.2 are not very precise, but they do force us to try to isolate the most important variables in the problem. We return to this theme in Section 5.6. Fermi questions also have proved useful for scientific discussion, especially when the available data are incomplete. Maybe the best known example is the Drake equation, which dates from the 1960s and has been used to motivate and justify the Search for Extraterrestrial Intelligence (SETI) project. Named for its originator, the American astronomer Frank Drake, the equation yields a rough estimate for the number N_c of extraterrestrial civilizations within the Milky Way galaxy that are technologically able to communicate with us. It is written as a product of factors, each of which can be estimated as

$$N_c = R \times f_p \times n_e \times f_\ell \times f_i \times f_c \times L. \tag{5.8}$$

In Equation 5.8, R is the annual rate of formation of suitable stars like our Sun in the Milky Way galaxy. Astronomical observations suggest that $R \approx 5$. The factor f_p is the fraction of those stars that have orbiting planets, n_e is the number of planets per star that are "earthlike" in their ability to support life, f_ℓ is the fraction of earthlike planets on which life develops, f_i is the fraction of the life-supporting planets where intelligent life evolves, f_c is the fraction of the planets with intelligent life where a viable interstellar communication technology such as radio is developed, and L is the lifespan of the advanced civilization, in years. The Drake equation can be considered to be a special case of Little's law, which describes flow balances. That law, named after the American economist John Little, also dates from the 1960s. It can be roughly summarized as stating that the number of items in a system (e.g., customers in a store) is equal to the average rate at which they exit the system, multiplied by the average time spent in the system. This simple concept can be applied to answer many Fermi questions.

The uncertainty in individual factors in the Drake equation is very large, especially for the factors f_ℓ, f_i, and f_c, which can range from zero to one. Consequently, estimates for N_c have a huge range, from less than 1 all the way up to 10,000. The Drake equation lies on the boundary between science and speculation, and it is up to the reader to determine how much faith, if any, to place in it. Speculating on the numerical value of N_c without data to support it does not really qualify as science. However, conventional areas of scientific investigation including astronomy, geology, biology, and planetary science and exploration can greatly reduce the uncertainty in factors such as f_ℓ appearing in the Drake equation.

Most estimates are based on some underlying balance. The estimate for the number of salons and barbershops in Example 5.1 is based on the economic principle that supply will meet demand in a free market economy. If there is a long waiting time to get a hair appointment, then more shops will inevitably open to fulfill the unmet need. If there are shops that are idle, some will close. Underlying the Drake equation is a balance between the birth and death of civilizations in our galaxy. As discussed in the next section, physical principles are particularly powerful for setting up these balances and thus very useful for making estimates.

5.3 ESTIMATES BASED ON SIMPLE PHYSICS

The physical sciences provide equations that allow us to quantitatively predict the outcome of experiments and, more broadly, describe events and conditions in the world. The resulting equations are often complicated and very difficult to solve. By combining simple physical principles with estimation techniques, however, reasonably accurate values can sometimes be obtained without finding an exact solution to the detailed equations. If the estimate is based on common sense and sound physical principles, then we can expect that the net result will at least have the correct order of magnitude, and often it is even closer. Such an estimate can also help determine whether a more rigorous calculation is justified and, if it is attempted, whether or not the calculation is on the right track.

Victor Weisskopf, an Austrian and American physicist, published some fascinating examples of such estimates in his book *Knowledge and Wonder* (Weisskopf 1992) and in a series of journal articles. In the next example, we have adapted Weisskopf's estimate for the maximum height that a mountain can attain on Earth.

EXAMPLE 5.3

Mount Everest is currently the tallest mountain on Earth. Its peak lies approximately 8848 m above sea level. Is there a maximal limit to the height of mountains on Earth, and is Mount Everest, with a height on the order of 10 km, near that limit?

ANSWER

A detailed analysis of this problem might involve trying to set up equations for mountain formation using plate tectonics and then searching for the onset of instabilities that could limit mountain growth. Alternatively, we will set up a very simple equation that only takes into account conservation of energy. The hope is that the simple equation will capture the main features of the problem and will provide an approximate numerical answer.

Following Weisskopf, we consider a mountain that already has attained its maximal height. Such a mountain must be supported against the force of gravity by the material under its base. If the mountain grows too tall, then its base can no longer support it, and the mountain will begin to sink under its own weight. This balance provides an idea for a simple solution to our problem.

As the mountain sinks, there is a release of gravitational energy. This energy release can be set equal to the energy required to liquefy the material under the base, which permits it to sink. We will further assume that the amount of material lost from the top of the mountain as it sinks approximately equals the amount of material caused to flow under the base. The net effect is equivalent to removing some material from the peak and replacing it at the base of the mountain. We can then equate the energy on a molecule-by-molecule basis. The release of gravitational energy as a single molecule is lowered from the peak of the mountain to the base is given by

$$U_{grav} = m_{molecule}gH, \tag{5.9}$$

where H is the height of the mountain, g is the acceleration due to gravity, and $m_{molecule}$ is the mass of a single molecule of the mountain, which we will assume is composed of quartz, i.e., silicon dioxide.

To estimate the energy required for the deformation at the base of the mountain, we consider the energy difference between the solid and liquid forms of quartz. This is the energy that it takes to melt quartz, which can be measured independently and is approximately 1.25×10^{-20} J/molecule. Equating the energy released from gravity to the energy absorbed for deformation yields

$$H \approx \frac{1.25 \times 10^{-20} \text{ J/molecule}}{m_{molecule}g}. \tag{5.10}$$

The other quantities on the right-hand side of Equation 5.10 can be found in standard tables. The mass of a silicon dioxide molecule is approximately

60 atomic mass units, or 1.0×10^{-25} kg, and $g = 9.8$ m/s^2. These values yield $H \approx 13$ km as our estimate for the upper limit on the height of mountains on Earth. A 100-km-tall mountain on Earth would literally melt under its own weight. The 13-km estimate is within a factor 2 of the actual height of Mount Everest, so the estimate is quite good. One reason for the slight overestimate might be that the rock at the base of the mountain only must be heated enough to become plastic so it can deform, which requires less energy than completely melting it.

Estimation techniques similar to Example 5.3 can also be applied to biophysics problems. The following problem is adapted from volume I of *The Feynman Lectures on Physics* (Feynman et al. 2005).

EXAMPLE 5.4

Animals such as some insects and crustaceans have compound eyes. Compound eyes do not have lenses to focus an image onto a retina, as do our own eyes, but instead are composed of many facets called ommatidia, each of which is an independent receptor of light. Each facet of the compound eye receives light from a limited direction and then acts like a tiny light pipe or fiber-optic cable. The net effect of the many facets is to enable the animal to discriminate the location of the object by forming an image, i.e., a rough map of the source of the light. Assuming that the radius of the entire compound eye of a large bee is 3 mm, estimate the optimal size d of a facet that produces the sharpest image.

ANSWER

In order for the bee to have the sharpest possible vision, we might think that its compound eye should have a huge number of facets, with each one collimating the light (i.e., guiding it in a straight line) to the center of the eye. The incoming light is collimated from a small angle

$$\theta_{\text{coll}} = \frac{d}{r}, \tag{5.11}$$

where r is the radius of the bee's eye and d is the facet diameter as schematically illustrated in Fig. 5.1. (An actual eye of a bee is not very spherical, but we will neglect that fact for this rough estimate.)

We often think that "more is better." If cost was no object, then we would tend to choose the digital display monitor with more lines of resolution or

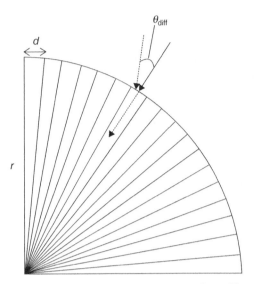

FIG. 5.1. Schematic representation of a bee's compound eye. The radius of the eye is r, and the diameter of each facet on the compound eye is d. Reducing the aperture d reduces the collimation angle $\theta_{coll} = d/r$, which improves the spatial resolution of the image. If d is too small, however, diffraction effects represented by θ_{diff} become important.

the digital camera with more megapixels, because these devices generally produce sharper images. We can imagine that even for these devices, however, progressively increasing the number of elements beyond some point will lead to diminishing returns, and eventually even become counterproductive. This is the case for the bee's compound eye.

Wave optics tells us that the incoming light is diffracted (i.e., bent, distorted, or broken up due to interference effects) when it passes through an aperture of diameter d. Diffraction effects become progressively important as d becomes comparable to or smaller than λ, the wavelength of the light. A ray of light* can enter straight into the facet, as shown by the solid arrow in Fig. 5.1. When diffraction effects are considered, however, a second ray of light can also enter the facet at a different angle, be bent, and then follow the same path as the first ray (dotted arrow in Fig. 5.1). The effect of diffraction is to confuse the directional discrimination of each facet and thereby blur the image that the bee perceives. When d is too small, diffraction effects dominate, which defeats the advantage of having many facets to improve the spatial resolution of the image. The trade-off can be quantified by using basic wave optics to estimate the angular spread from diffraction:

*Strictly speaking, a "ray" of light is a concept from geometric optics, where diffraction effects are assumed to be negligible. Here we use the term more loosely, i.e., a line perpendicular to the wave fronts.

$$\theta_{\text{diff}} \approx \frac{\lambda}{d}. \tag{5.12}$$

We could refine Equation 5.12 by including a multiplicative factor, on the order of 1, from optics if we know the shape of the facet, but again we will neglect these details. Equations 5.11 and 5.12 together provide the balance or trade-off, which is the basis of this estimate. If d is too large, then according to Equation 5.11, the incoming light is not well collimated, and the image will be blurry. If d is too small, then diffraction effects dominate, and again the image will be blurry. The net angular spread of incoming light that is accepted into a facet is determined by a combination of both effects and is approximately

$$\theta_{\text{net}}(d) = \theta_{\text{coll}} + \theta_{\text{diff}} = \frac{d}{r} + \frac{\lambda}{d}. \tag{5.13}$$

The sharpest image that the bee can perceive results when θ_{net} is minimized. Setting the derivative of θ_{net} with respect to d equal to zero (and then checking that the second derivative is positive) yields the optimal values $d_{\text{opt}} = \sqrt{\lambda r}$ and $\theta_{\text{net,opt}} = 2\sqrt{\lambda/r}$. Substituting $\lambda = 475\,\text{nm}$ (i.e., the wavelength of blue visible light) and $r = 3\,\text{mm}$ yields a rough estimate of $d_{\text{opt}} \approx 38 \times 10^{-6}\,\text{m}$ and $\theta_{\text{net,opt}} \approx 0.025\,\text{rad}$ or $1.4°$. This estimate for the aperture d of a large bee's ommatidium is accurate to within a factor of 2 of the values found in nature ($20 - 30 \times 10^{-6}\,\text{m}$).

The estimated value for θ_{net} also indicates that the bee's vision is not as sharp as our own. For example, a full moon subtends approximately $0.5°$ in the sky, or about a third of the estimate for θ_{net}, so a bee cannot see any detail on an object of that angular extent. (A $0.5°$ angular extent is also roughly equal to a thumbnail's width at an arm's length, for most people.) Instead, the estimate suggests that a bee perceives the image of an object subtending $0.5°$ as highly blurred. Equation 5.13 indicates that this blurry vision is not due to an inefficient design of the bee's visual system, but instead is a limitation based on its eye's small size relative to the wavelength of visible light.

5.4 APPROXIMATING DEFINITE INTEGRALS*

Many processes that are modeled by equations involve continuous variation and are best described using calculus. Whenever we want to total up how much of a particular quantity accumulates over the course of a process, it is usually necessary to evaluate a definite integral describing the duration of the process.

*The material presented in this section following Example 5.5 (and associated exercises 5.5 and 5.13) assumes familiarity with complex analysis and can be skipped by those who have not yet studied that subject.

Sometimes, we can get the result exactly. By way of illustration, consider the variation over time of the population $p(t)$ of a biologic organism as modeled with the differential equation

$$\frac{dp(t)}{dt} = K p(t) - A p^2(t). \tag{5.14}$$

The function $p(t)$ could describe the number of bacteria in a Petri dish. When $A = 0$, Equation 5.14 is straightforward to solve. We obtain simple exponential growth, so we do not need to analyze that special case in detail. The positive constant K describes the population growth rate, while the positive constant A describes factors limiting the population growth, such as overcrowding or a limited food supply. By setting the left-hand side of Equation 5.14 to zero, we obtain the steady-state population $p_{ss} = K/A$. If, however, the population has not yet reached its steady-state value, then a more complete solution is necessary. Equation 5.14 is an example of a separable differential equation, so one of the steps to solve for $p(t)$ is to calculate the definite integral

$$t - t_0 = \int_{u=p_0}^{p(t)} \frac{du}{\left(Ku - Au^2\right)}, \tag{5.15}$$

where p_0 is the initial value of the population at $t = t_0$ and u is an integration variable.

There are several methods commonly used to evaluate definite integrals. If we are fortunate, an antiderivative for the corresponding indefinite integral is listed in a table or can be evaluated with symbolic manipulation software. We can solve Equation 5.15 for $p(t)$ by looking in a standard integral table, where we find

$$\int \frac{dx}{bx + ax^2} = \frac{1}{b}\ln\left(\frac{x}{ax+b}\right) + \text{constant}. \tag{5.16}$$

Applying Equation 5.16 to Equation 5.15 with the substitutions $x = u$, $a = -A$, and $b = K$ yields

$$t - t_0 = \frac{1}{K}\left(\ln\frac{p(t)}{K - Ap(t)} - \ln\frac{p_0}{K - Ap_0}\right). \tag{5.17}$$

After some algebra, we can solve exactly for $p(t)$:

$$p(t) = \frac{p_0 e^{K(t-t_0)}}{1 + \left(\dfrac{p_0 A}{K}\right) \times \left(e^{K(t-t_0)} - 1\right)}. \tag{5.18}$$

Note that Equation 5.18 is consistent with the initial value $p(t)|_{t=t_0} = p_0$ and also the steady-state value $p_{ss} = p(t)|_{t \to \infty} = K/A$. Also, when $A = 0$, Equation 5.18 reduces to simple exponential growth, as expected. By fitting Equation 5.18 to observed data (population measurements at multiple time points), the values of the constants A and K can be extracted. How well Equation 5.18 describes the observed data can also determine whether the model in Equation 5.14 needs to be refined. We return to the modeling of population in Chapter 8 to study a related difference equation that illustrates some important and surprising elements of chaos theory.

Unfortunately, there are many indefinite integrals for which a simple result like Equation 5.16 cannot be written. In these cases, approximation methods are needed. For example, the results for even relatively innocuous-looking indefinite integrals such as

$$\int \sqrt{1 + x^3}\, dx \quad \text{or} \quad \int e^{-x^2}\, dx \tag{5.19}$$

cannot be expressed in terms of "elementary" functions such as polynomials, sines, cosines, exponentials, and logarithms.

Even if an indefinite integral cannot be found in tables, a numerical method can be used to evaluate the definite integral. Many numerical integration techniques have been developed, and they are very practical and widely used. Popular methods include the trapezoidal approximation, Simpson's method, Gaussian integration, Monte Carlo methods, and others. With Simpson's approximation method, for example, the integrand is approximated by a series of contiguous, equally spaced parabolic segments, which are straightforward to integrate. The accuracy of the approximation increases with the number of segments.

Using numerical methods, we can calculate a definite integral related to one of the indefinite integrals listed in Equation 5.19:

$$\int_{-\infty}^{\infty} e^{-u^2}\, du = 1.77245\ldots = \sqrt{\pi}. \tag{5.20}$$

The definite integral in Equation 5.20 can also be evaluated analytically (see exercise 5.6). Using variable substitution or the method based on the dimensional checks described in Chapter 1, Equation 5.20 can be generalized to

$$\int_{-\infty}^{\infty} e^{-au^2}\, du = \sqrt{\frac{\pi}{a}}. \tag{5.21}$$

Although numerical integration methods are very widely used, they are not the focus of the current discussion. When using numerical integration, the bulk of the work goes on inside the computer and consequently is usually not expressed by equations. Instead, we explore several analytic methods that are

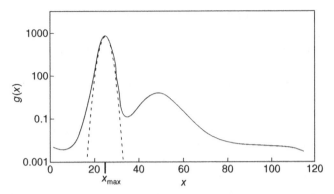

FIG. 5.2. A function $g(x)$ with a dominant maximum at $x = x_{max}$. Its peak can be fitted with a Gaussian function, which when plotted on a logarithmic scale, appears as an inverted parabola (dashed line).

used to approximate definite integrals. One advantage of analytic approximation methods is that they allow us to retain symbolic constants and parameters all the way through to the final result.

We discuss three closely related methods that provide approximate values for definite integrals when the major contribution comes from a limited region. The first method we discuss is known as *Laplace's method* (named after Pierre-Simon Laplace, an eighteenth- and nineteenth-century French mathematician and astronomer). Laplace's method assumes that the integrand has a dominant maximum. When the integrand is plotted on a linear scale, the maximum might look like a sharp spike. By plotting the integrand on a logarithmic scale as in Fig. 5.2, however, more of the structure of the maximum is apparent. We can fit a Gaussian function (which appears as an inverted parabola on the log scale) to the integrand. The area under the Gaussian can be readily calculated, which allows us to estimate the integral.

Consider the general definite integral

$$I = \int_{x_1}^{x_2} e^{g(x)} dx, \tag{5.22}$$

where the function $g(x)$ is a real function that has a dominant maximum at x_{max}, lying within the region of integration $x_1 \leq x \leq x_2$. A Taylor series expansion of $g(x)$ about x_{max} yields

$$g(x) = g(x_{max}) + (x - x_{max}) \frac{dg}{dx}\bigg|_{x=x_{max}} + \frac{1}{2!}(x - x_{max})^2 \frac{d^2 g}{dx^2}\bigg|_{x=x_{max}} + \cdots. \tag{5.23}$$

The series is named after the English mathematician Brook Taylor, who published his work in 1714. Because g has a maximum at x_{max}, its first derivative is zero at that point, and its second derivative is negative, so ignoring higher-order terms yields

$$I \approx e^{g(x_{max})} \int_{x_1}^{x_2} e^{-\frac{1}{2}\left|\frac{d^2g}{dx^2}\right|(x-x_{max})^2} dx. \tag{5.24}$$

Because we are assuming that the dominant contribution to the integral I is from the vicinity of $x = x_{max}$, it is a reasonable approximation to extend the lower limit of integration from x_1 to $-\infty$, and the upper limit from x_2 to $+\infty$. Applying Equation 5.21 with $u = (x - x_{max})$ yields the desired result for Laplace's method:

$$I \approx e^{g(x_{max})} \times \sqrt{\frac{2\pi}{\left|\frac{d^2g}{dx^2}\right|_{x=x_{max}}}} = e^{g(x_{max})} \times \sqrt{\frac{2\pi}{|g''(x_{max})|}}. \tag{5.25}$$

Laplace's method provides a good estimate for I whenever the largest values of the integrand are limited to a small region along the integration path. Laplace's method is illustrated by the well-known example of calculating Stirling's approximation for the factorial function $n!$, for large n. It is named after James Stirling, an eighteenth-century Scottish mathematician.

EXAMPLE 5.5

The factorial function $n! = 1 \times 2 \times ... \times n$ (with $0! = 1$) increases monotonically and very rapidly with n. When $n \sim 70$, the value of $n!$ exceeds 10^{100} (i.e., one googol), which will cause an overflow fault on most scientific calculators. This makes factorials of large numbers difficult to work with.

(a) Show that the factorial can be represented by the definite integral

$$I_n = \int_0^\infty t^n e^{-t} dt = n! \tag{5.26}$$

for $n = 0, 1, 2, ...$

(b) Use a variable substitution and apply Laplace's method to verify Stirling's approximation

$$n! \approx \sqrt{2\pi n}\, n^n e^{-n} \tag{5.27}$$

for large values of n. Equation 5.27 is the first term in an asymptotic series, that is, the ratio of the two sides approaches 1 as $n \to \infty$.

ANSWER

(a) When $n = 0$, the integral is straightforward to evaluate, yielding $I_0 = 1$ in agreement with the definition $0! = 1$. Integration by parts can be applied to the related indefinite integral

$$\int t^n e^{-t} dt = -t^n e^{-t} + n \int t^{n-1} e^{-t} dt. \tag{5.28}$$

Note that integration by parts reduces the exponent of t in the integrand by one unit. When n is an integer greater than zero, this process can be iterated n times to yield (up to an additive integration constant)

$$\int t^n e^{-t} dt = -e^{-t}\left(t^n + nt^{n-1} + n(n-1)t^{n-2} + n(n-1)(n-2)t^{n-3} + \dots + n!\right). \tag{5.29}$$

Only the last term survives when Equation 5.29 is evaluated at infinity and zero, verifying Equation 5.26. Alternatively, Equation 5.26 can be verified by mathematical induction.

(b) Let $x = t/n$ and recognize that $x^n = e^{n\ln x}$, so that Equation 5.26 becomes

$$I_n = n^{n+1} \int_0^\infty e^{g(x)} dx, \tag{5.30}$$

where

$$g(x) = n(\ln x - x). \tag{5.31}$$

The function $g(x)$ attains a maximum of $g(x_{max}) = -n$ when $x_{max} = 1$. For large values of n, this maximum is amplified, justifying the use of Laplace's method. Substituting the values $g(x_{max})$ and the absolute value of the second derivative $|g''(x_{max})| = n$ into Equations 5.25 and 5.30 yields Equation 5.27.

A definite integral is evaluated by following a specific integration path. In most cases, only the end points of the paths are specified, so there is a great deal of flexibility in choosing the precise path. This is particularly true if the integration path is allowed to leave the real axis and traverse elsewhere into the complex plane. Even if the end points of the definite integral both lie on the real axis, there is no obligation to choose a path that lies solely along the real axis. By choosing a path in the complex plane, we might be able to find a dominant maximum of the integrand and use it to approximate the integral. When Laplace's method is generalized to integration paths in the complex plane, it is known as the *method of steepest descents*.

Cauchy's integral theorem (named after Augustin Louis Cauchy, a French mathematician, who published this work in 1814) provides the mathematical

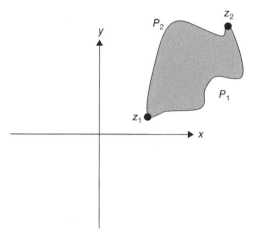

FIG. 5.3. Provided the function $f(z)$ has no singularities in the shaded region or its boundary, Cauchy's integral theorem ensures that integration over paths P_1 and P_2 yields the same result.

justification to change the integration path. The theorem states that if a function is analytic within a region and on its boundary, then the closed-loop integration along the boundary is zero:

$$\oint f(z)\,dz = 0. \tag{5.32}$$

Recall that "analytic" means that the derivative of $f(z)$ exists, which implies that there are no singularities in the region. Functions can have several types of singularities in the complex plane, the most common of which include poles, i.e., infinities such as $f(z) = (z-5)^{-1}$ at $z = 5$, and branch points, i.e., multiple values, such as $f(z) = (z-7)^{1/2}$ at $z = 7$. A more complete discussion is provided in Gamelin (2003).

Cauchy's integral theorem means that we are free to warp the integration path in the complex plane as in Fig. 5.3, provided we do not cross any singularities. In particular, the integrals over paths P_1 and P_2 must be equal because

$$\oint f(z)\,dz = \int_{P_1} f(z)\,dz - \int_{P_2} f(z)\,dz = 0. \tag{5.33}$$

EXAMPLE 5.6

Evaluate the integral

$$\int_{-\infty}^{\infty} e^{iax^2}\,dx = \int_{-\infty}^{\infty} \cos ax^2\,dx + i \int_{-\infty}^{\infty} \sin ax^2\,dx. \tag{5.34}$$

(where a is a real, positive constant) by warping the integration path away from the real axis.

ANSWER

The limits of integration lie on the real axis, where the magnitude of the integrand

$$\left|e^{iax^2}\right| = \sqrt{\cos^2(ax^2) + \sin^2(ax^2)} = 1 \qquad (5.35)$$

for all values of x. Although we could integrate along the real axis, this path gives no hint of a dominant contribution needed for the successful application of the method of steepest descents. We will consider that integration path again, later in this section.

Let us examine the integrand as a function of the complex variable $z = x + iy$, so that $z^2 = x^2 - y^2 + 2ixy$ and

$$e^{iaz^2} = e^{ia(x^2 - y^2)} e^{-2axy}. \qquad (5.36)$$

From Equation 5.36, we see that e^{iaz^2} has magnitude and phase

$$\left|e^{iaz^2}\right| = e^{-2axy} \quad \text{and} \quad \measuredangle e^{iaz^2} = a(x^2 - y^2). \qquad (5.37)$$

The derivative of e^{iaz^2} is $2iaze^{iaz^2}$, which vanishes when $z = 0$. We will see that depending on the integration path, i.e., the specific values of x and y traversed in the complex plane, $z = 0$ can be either a minimum or a maximum for $\left|e^{iaz^2}\right|$. It can also be neither a max nor a min, as is apparent from Equation 5.35. Because of its min/max nature, the origin $z = 0$ is called a *saddle point* for this integrand. The reason behind this name is apparent from the graph in Fig. 5.4.

Consider a straight-line path through $z = 0$ oriented $+45^0$ from the real axis: $z = s \times e^{i\pi/4} = s\sqrt{i}$. Along the path $x = y$, the real parameter s measures progress along the line. Following this path, the value of integrand e^{iaz^2} is e^{-as^2}, which has a maximum at $s = 0$. If instead we follow the path along the -45^0 line, then $z = s \times e^{-i\pi/4}$, the path is described by $x = -y$, and the value of the integrand is e^{+as^2}, which has a minimum at $s = 0$. These results explicitly illustrate the saddle-point nature of e^{iaz^2} at $z = 0$.

To capitalize on the maximum, we can evaluate the integral using the $+45^0$ path through the origin (i.e., $z = s \times e^{i\pi/4}$). Such a path can be obtained by warping the original path along the real axis into the one shown in Fig. 5.5. The value of the definite integral is unchanged by this procedure because no singularities are crossed. Note that along the two vertical segments of the warped path, the sign of the product xy is always positive, so the factor e^{-axy} in Equation 5.36 ensures that the contribution to the integral from

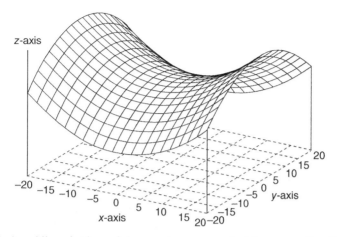

FIG. 5.4. A saddle point is a minimum when approached from one direction and a maximum when approached from a different direction.

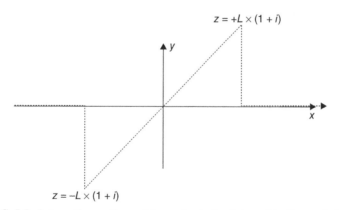

FIG. 5.5. Integration path used to evaluate the integral in Example 5.6.

those two segments is suppressed. As $L \to \infty$, the only surviving contribution to the integral is from the diagonal line. Because, in this case, the integral is already in the form of a Gaussian, the expansion of Equation 5.23 is unnecessary. Therefore, the exact result is

$$\int_{x=-\infty}^{\infty} e^{iax^2} dx = e^{i\pi/4} \int_{s=-\infty}^{\infty} e^{-as^2} ds = e^{i\pi/4} \sqrt{\frac{\pi}{a}} = (1+i)\sqrt{\frac{\pi}{2a}}. \qquad (5.38)$$

Choosing the path that passes over the maximum at the saddle point has provided the value of the two definite integrals on the right-hand side of Equation 5.34. Equating the real and imaginary parts of Equation 5.38 yields

$$\int_{-\infty}^{\infty} \cos ax^2 dx = \sqrt{\frac{\pi}{2a}} \qquad\qquad (5.39)$$

and

$$\int_{-\infty}^{\infty} \sin ax^2 dx = \sqrt{\frac{\pi}{2a}}. \qquad\qquad (5.40)$$

If instead we were to follow an integration path along the real axis by constraining the value of y to be 0, then from Equation 5.37, the magnitude of the integrand is always 1 and its phase varies quadratically with x. The oscillation from this rapidly varying phase provides a great deal of cancellation. The cancellation is least near $x = 0$, however, where the frequency of oscillation is slowest (Fig. 5.6). From this alternative point of view, we are also led to expect the largest contribution to the integral from the region near the origin. This reasoning motivates the third, related technique to evaluate definite integrals, known as the method of *stationary phase*. We will consider a more general form of the integrand given by $e^{ih(x)}$, where $h(x)$ is a real-valued function that has a dominant *minimum* at x_{min}.

We can then follow steps analogous to those used to derive Laplace's method, but now seek a minimum rather than a maximum. Recalling that the second derivative of $h(x)$ is positive at the minimum, the result is

$$I = \int_{x_1}^{x_2} e^{ih(x)} dx \approx e^{ih(x_{min})} \times \sqrt{\frac{2\pi i}{h''(x_{min})}} = e^{ih(x_{min})} \times \sqrt{\frac{\pi}{h''(x_{min})}} \times (1+i). \quad (5.41)$$

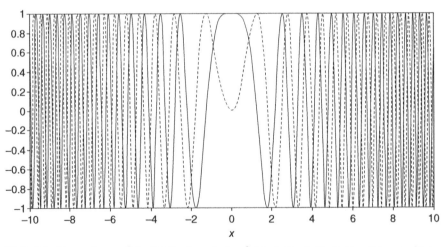

FIG. 5.6. Plots of $\cos x^2$ (solid line) and $\sin x^2$ (dash line). Both functions oscillate rapidly everywhere except in the neighborhood of $x = 0$, illustrating the method of stationary phase.

Notice that when $h(x) = ax^2$ and the limits of integration are $-\infty$ to $+\infty$ as in Equation 5.38, $h''(x_{min}) = 2a$, so that we recover the result of Example 5.6. Specifically, the stationary phase result given in Equation 5.41 is the same as Equation 5.38. Both the methods of steepest descent and stationary phase select the same location for the dominant contribution to the definite integral in the saddle-point region. Although the two involve taking different integration paths, both are examples of *saddle-point* methods.

Saddle-point integration methods were used to great advantage by the twentieth-century German–American physicist Hans Bethe. In his work in nuclear theory, Bethe started with an expression from statistical mechanics involving the two-dimensional integral

$$Z(\alpha,\beta) = \int_{A=-\infty}^{\infty} \int_{E=-\infty}^{\infty} \rho(A,E)e^{(\alpha A - \beta E)}dAdE, \tag{5.42}$$

where $\rho(E,A)$ is the density of levels (number of levels per energy interval) at energy E for a system consisting of A particles. In statistical mechanics, the function Z is known as the "grand partition function." His aim was to learn more about the density function $\rho(A,E)$, but this quantity is often difficult to determine. At least formally, however, it can be expressed by inverting Equation 5.42 using a two-dimensional inverse Laplace transform:

$$\rho(A,E) = \frac{1}{(2\pi i)^2} \int_{\alpha=-i\infty}^{+i\infty} \int_{\beta=-i\infty}^{+i\infty} Z(\alpha,\beta)e^{-(\alpha A - \beta E)}d\alpha d\beta. \tag{5.43}$$

As explained in texts on Laplace transforms, Equation 5.43 assumes that all the singularities of $Z(\alpha,\beta)$ occur at points with $Re(\alpha) < 0$ and $Re(\beta) < 0$. We can then apply saddle-point approximation to evaluate the integral in Equation 5.43 for large values of α and β. This method is outlined in exercise 5.13. Bethe's result is remarkable. Initially, it seems that it will be impossible to extract any information about the density function $\rho(A,E)$ because it is embedded in the integrand of Equation 5.42.

Sometimes, these approximation methods for definite integrals not only provide us with a numerical answer but also enhance our qualitative understanding of the solution. One example is the insight into the analysis of Fermat's principle of least time. In the seventeenth century, the French mathematician Pierre de Fermat postulated that when light travels from point A to point B, it always traverses the path that requires the smallest possible elapsed time. This is certainly expected for light traveling in a straight beam, but it is quite a remarkable result when the light is bent or refracted, e.g., at the air–water interface of a pond. Experiments have shown that Fermat's principle of least time is correct.

The interpretation of Fermat's principle of least time grew even more perplexing in the early twentieth century, when it became apparent that light is composed of particles, or quanta, known as photons. How do the photons "know" to travel in the exact path that yields the minimum elapsed time? An explanation was provided later in the century by a formulation of quantum mechanics developed by Richard Feynman and others. In that theory, the probability for the photon to take a specific path is expressed by a definite integral. Most of the paths produce negligible probability due to the oscillatory nature of the integrand. The path corresponding to least time produces a stationary phase and thus yields a higher probability.

Feynman's work was also mentioned in connection with the rough estimate related to the bee's eye described in Example 5.4. Scientists like Fermi, Feynman, and Weisskopf mentioned in this chapter displayed remarkable skill at estimation and approximation while also excelling at rigorous mathematical calculation. It is not a coincidence that those luminaries of twentieth-century physics appreciated the power of qualitative and semiquantitative estimates and continually strove to hone their skills. In this century, computers have become increasingly proficient at algebra and other symbolic manipulation. The value of estimation skills, which so far only humans have mastered, has only increased.

5.5 PERTURBATION ANALYSIS

The problems that emerge in science and engineering often lead to differential equations, or related expressions involving difference or matrix equations. Because only a handful of these resulting equations have known, analytic solutions, a variety of approximation methods have been devised. One class of methods can be applied when the problem of interest is "close" to one of the cases that can be solved exactly. These techniques are called *perturbation* methods, because the desired problem can be reached from the solvable case by a small change, or perturbation.

Perturbation methods have been developed for general, mathematical problems that are not specifically associated with any specific physical situation. For example, a standard perturbation problem is: given a symmetric, $N \times N$ matrix $\mathbf{A}$ with known eigenvalues $\lambda_1, \lambda_2, \ldots, \lambda_N$ and known eigenvectors $\vec{v}_1, \vec{v}_2, \ldots, \vec{v}_N$, then find the eigenvalues and eigenvectors for the matrix $\mathbf{A}'$,

$$\mathbf{A}' = \mathbf{A} + \mathbf{B}\varepsilon, \tag{5.44}$$

where $\mathbf{B}$ is another symmetric matrix and $\varepsilon \ll 1$. The term $\mathbf{B}\varepsilon$ represents a small perturbation to the matrix $\mathbf{A}$. (Eigenvalues of a symmetric matrix were discussed in connection with Example 4.3, where a special case amenable to a graphical solution was considered.)

We will not delve into the details here, but instead only mention that there is a known algorithm to solve this problem. The main feature of the solution is to assume that the eigenvalues of $\mathbf{A}'$ can be expressed as a series expansion in powers of ε. For example, for the first eigenvalue of $\mathbf{A}'$, we write

$$\lambda_1' = \lambda_1 + \varepsilon c_{11} + \varepsilon^2 c_{12} + \varepsilon^3 c_{13} + \dots \qquad (5.45)$$

Similarly, the first eigenvector of $\mathbf{A}'$ can be expressed in terms of the first eigenvector of $\mathbf{A}$, plus correction terms with increasing powers of ε:

$$\vec{v}_1' = \vec{v}_1 + \varepsilon \vec{v}_{11} + \varepsilon^2 \vec{v}_{12} + \varepsilon^3 \vec{v}_{13} + \dots \qquad (5.46)$$

For sufficiently small values of ε, the series in Equations 5.45 and 5.46 are expected to converge. The hope is that the first two (i.e., up to linear in ε) or possibly three terms will give good approximations for the true results for λ_1' and $\vec{v}_1'$. By requiring that the difference $(\vec{v}_1' - \vec{v}_1)$ is orthogonal to $\vec{v}_1$ and by requiring that $\mathbf{A}'\vec{v}_1' = \lambda_1'\vec{v}_1'$ holds for each power of ε, the coefficients c_{ij} and $\vec{v}_{ij}$ can be found in terms of the perturbing matrix $\mathbf{B}$ and the known eigenvalues and eigenvectors of the matrix $\mathbf{A}$. We then repeat the process for λ_2' the second eigenvalue of $\mathbf{A}'$ and so on. The details of the algorithm are discussed in standard texts (e.g., Schiff 1968 or Davydov 1976).

Many times perturbation methods are linked with a specific physical problem. Perhaps, the best known example is the orbit of a planet. The two-body problem (e.g., the Earth orbiting the Sun) can be solved analytically using Newtonian mechanics. However, when a third, perturbing body is added (e.g., the Moon and its weak gravitational pull on the Earth), we can no longer find a general, analytic solution. Today, numerical methods can be used to solve for the resulting orbit of the Earth, but for the nearly 300-year gap between the publication of Newton's *Principia Mathematica* and the advent of electronic computers, perturbation methods were the most effective option available.

Perturbation methods can be applied not only to problems in classical mechanics like the celestial orbits but also to problems in quantum mechanics that describe submicroscopic particles like atoms and molecules. We tend to think of quantum mechanical calculations as being more difficult, possibly because classical mechanics deals with objects on our own size scale. Oddly, perturbation methods are usually much easier to apply to quantum mechanical problems than to their classical analogs. One reason for this surprising result is that while a classical perturbation, like the gravitational pull of the Moon, is small in comparison with the Sun's pull, its cumulative effect on the Earth's orbit over the course of time can be quite large. In quantum mechanics, we can often avoid this difficulty by finding so-called stationary states, which are solutions to the Schrödinger equation associated with a specific value of the energy E_n. The time dependence in the stationary states can be separated out into a multiplicative factor:

$$\psi(x,t) = \psi_n(x) \times e^{i2\pi E_n t/h}, \qquad (5.47)$$

where h is Planck's constant. The cumulative effect of the perturbation over time is not a concern for time-independent factor $\psi_n(x)$ in Equation 5.47. Provided that the perturbation itself is time independent, then small perturbations often can be expressed in terms of a rapidly converging series. Also, the equations of quantum mechanics can be expressed in terms of matrices, so the same mathematical algorithms that were described in connection with Equations 5.44 and 5.45 can be applied directly. We will describe an example from physical chemistry and materials science that illustrates how quantum mechanical perturbation methods are applied. We will skip over many of the mathematical details, which are relatively complicated, and instead present an overview.

Many problems in mechanics involve particles subject to the influence of a local potential energy function U that depends on position. The spatial derivative of this potential energy function yields the force at each location, which, in the classical mechanical description, causes the particles to accelerate. If U is a function of the single variable r (e.g., the problem is one-dimensional or else has radial symmetry), then

$$F(r) = -\frac{dU(r)}{dr} = ma. \tag{5.48}$$

In quantum mechanics, the potential energy function directly governs the motion, and forces are not used directly in the calculation. Suppose $U(r)$ has a minimum at $r = r_{min}$. Then we can expand the potential energy function in a Taylor series:

$$U(r) = U(r_{min}) + (r - r_{min})U'(r_{min}) + \frac{1}{2!}(r - r_{min})^2 U''(r_{min})$$

$$+ \frac{1}{3!}(r - r_{min})^3 U'''(r_{min}) + \dots. \tag{5.49}$$

The first term $U(r_{min})$ is a constant. This constant term has a zero derivative, so classically, it does not produce any force. We can think of $U(r_{min})$ as an arbitrary constant, because only the difference in potential energy between two locations r_1 and r_2 is a physically important quantity. The linear term vanishes because the first derivative is zero at the location of the minimum. The first nonzero, physically important term is the quadratic or parabolic term. For sufficiently small values of $x = r - r_{min}$, the cubic and higher-order terms in Equation 5.49 are negligible.

One familiar example of this parabolic term is the potential energy function for a mass on a spring. The connection is easier to see if we write

$$U(r) - U(r_{min}) \approx \frac{1}{2!}(r - r_{min})^2 U''(r_{min}) = \frac{1}{2}kx^2. \tag{5.50}$$

We have associated the spring constant k with the second derivative, $k = U''(r_{min})$ and x as the displacement from the minimum, $x = (r - r_{min})$. Recall that a mass m attached to a spring with constant k acts as a harmonic oscillator. The frequency f (in hertz) of the oscillation is given by

$$f = \frac{1}{2\pi}\sqrt{\frac{k}{m}}. \tag{5.51}$$

The parabolic potential energy function is very convenient to work with. In Section 5.4, we discussed some of the advantages of working with the quadratic term in a series expansion in connection with Laplace's approximation for definite integrals. There are well-known solutions to the harmonic oscillator problem, both in classical and quantum mechanics. We will start with the known solution to the quantum mechanical version of this problem and use it as the starting point for the perturbation analysis.

The quantum mechanical solution for the harmonic oscillator is characterized by a series of discrete energy levels characterized by the index or "quantum number" n

$$E_n = hf\left(n + \frac{1}{2}\right), \quad n = 0, 1, 2, \ldots, \tag{5.52}$$

where h is Planck's constant. From Equation 5.52, the energy gap between two adjacent levels of the quantum harmonic oscillator is

$$\Delta E = E_{n+1} - E_n = hf \tag{5.53}$$

independent of n.

We will try to apply the harmonic oscillator model to the atoms in a bar of iron. Consider the potential energy function associated with the force between two adjacent atoms in the lattice of iron atoms. We observe that a bar of iron is stable; it neither falls apart nor collapses in on itself. We can therefore infer that the potential energy function between any two iron atoms has a minimum, because it is unfavorable (in terms of energy) for the two atoms to approach each other too closely or stray too far apart from each other. Because of the existence of the minimum, Equations 5.49 and 5.50 suggest that the two iron atoms oscillate about an equilibrium separation r_{min}. That is, to a lowest-order approximation, the two atoms will vibrate as if there were a tiny spring joining them together, and their relative separation r will undergo simple harmonic motion. Because of the small scale of the problem, quantum mechanics is the appropriate method to analyze the two attached atoms.

Experimental observations support this simple model. Solid iron emits a very small but detectable amount of electromagnetic radiation in the infrared frequency range. The spectrum of this electromagnetic radiation is consistent with the model of relative vibrational motion of the two adjacent iron atoms. If the harmonic oscillator makes a transition between quantum states $n + 1$ and n, then from Equation 5.53, $f = \Delta E/h$ is the frequency of the emitted electromagnetic radiation. This frequency can be measured, so using Equation 5.51, we can determine the spring constant k if we know the mass of an iron atom! Note that because r is the relative separation of the two atoms, it turns out that we have to use the "reduced" mass of the pair of iron atoms for m in Equation 5.51, which is one-half the mass of each iron atom. This minor complication is not central to this discussion and is explained in most books on classical mechanics.

The harmonic oscillator model for the iron lattice appears to be a success, but if we look at more experimental data, we begin to see problems. This is not surprising for such a simplified model. The parabolic potential energy function alone cannot explain the further observations that the bar of iron expands very slightly when heated, and also that the value of the coefficient that describes the expansion itself depends on temperature.

We know that a harmonic oscillator, like a mass on a spring or like a child on a swing, spends an equal amount of time on either side of the equilibrium point. As we heat up the iron, we are adding thermal energy to the atoms in the lattice. This is analogous to pushing the child on the swing higher and higher. No matter how much energy we add to the harmonic oscillator, however, the average value of $x = r - r_{min}$ over one period remains zero. This can be shown mathematically from the quantum mechanical solution for the harmonic oscillator but is also apparent from the symmetry about the line $r = r_{min}$ of the parabolic potential energy function in Equation 5.50. As a result, the average value of separation of the iron atoms remains r_{min} in the harmonic oscillator model, independent of the temperature of the bar of iron. The observation that the bar of iron expands requires that there be a slight increase in the average atomic separation, and the simple harmonic oscillator model does not predict this.

Empirically, the thermal expansion is described by a formula for the length L of the bar of iron as a function of temperature T relative to some reference temperature, which is usually taken to be near room temperature, e.g., $T_0 = 20\,°C$:

$$\frac{L(T) - L(T_0)}{L(T_0)} = \alpha \times (T - T_0). \tag{5.54}$$

In Equation 5.54, α is the linear thermal expansion coefficient in unit $(°C)^{-1}$. Measurements show that the expansion coefficient α is a very small, positive value on the order of $1 \times 10^{-5}\,°C^{-1}$. This small value suggests that perturbation methods could be applicable to this problem, i.e., the observed physical situation is close to the previously solved problem of the harmonic oscillator, which predicts $\alpha = 0$.

We return to the original full potential energy function $U(r)$ in Equation 5.49, which we have approximated in the vicinity of its minimum by Equation

5.50. This approximation is not valid when the separation of the two iron atoms differs greatly from r_{min}. The full solution of the relative motion of the two iron atoms will be more difficult to find for the actual potential energy function than for its parabolic approximation. The strategy will be to assume a more realistic form for the potential energy function, then expand it up to third order in powers of $(r - r_{min})$. The third-order term can be treated as a small perturbation to the harmonic oscillator problem.

EXAMPLE 5.7

A simple, but relatively realistic model for the interatomic potential energy function is

$$U_M(r) = A\left(e^{-2a(r-r_0)} - 2e^{-a(r-r_0)}\right),\qquad(5.55)$$

known as the generalized Morse potential after Philip M. Morse, a twentieth-century American physicist. The Morse potential is a good model for the interatomic forces in a molecule, although its validity begins to break down when $r \ll r_0$. Ironically for our example, as described in Chen (2004), the Schrödinger equation with a Morse potential energy function can be solved analytically, but the results are quite messy, so we will use this function to illustrate perturbation methods anyway.

(a) Find the minimum of $U_M(r)$ and expand the potential energy function in Equation 5.55 up to fourth order. Interpret the physical meaning of the three free parameters r_0, a, and A in Equation 5.55.

(b) Estimate the ratio of the cubic to quadratic term in the expansion of $U_M(r)$ based on the empirical values for A and $\Delta E = hf$. Is this problem a good candidate for perturbation methods?

ANSWER

The minimum occurs where the derivative is zero:

$$\frac{dU_M}{dr} = -Aa\left(2e^{-2a(r-r_0)} - 2e^{-a(r-r_0)}\right) = 0.\qquad(5.56)$$

By inspection, we see that Equation 5.56 is satisfied when $r = r_0$. We can readily verify that the second derivative is positive at that location, so r_0 is

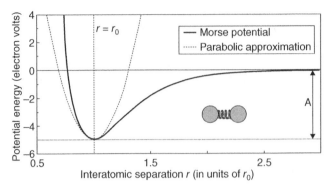

FIG. 5.7. Interatomic forces can be modeled with the Morse potential energy function in Equation 5.56. The function has a minimum at $r = r_0$ and can be approximated by a parabola.

the location of a minimum r_{min}. Then Morse potential can be expanded in powers of $r - r_0$ about its minimum to give

$$U(r) = -A + Aa^2(r-r_0)^2 - Aa^3(r-r_0)^3 + \frac{14}{24}Aa^4(r-r_0)^4 + \dots. \quad (5.57)$$

The meaning of the parameter A can be interpreted from Equation 5.56 and Fig. 5.7:

$$U_M(\infty) - U_M(r_{min}) = +A. \quad (5.58)$$

The parameter A represents the binding or disassociation energy, i.e., the amount of energy required to pull apart the pair of atoms and separate them to $r = \infty$. Finally, from Equation 5.55, the parameter a controls the width of the energy well of the Morse potential plotted in Fig. 5.7. Increasing a increases the steepness of the well. Comparing Equations 5.50 and 5.57, we also see that a is related to spring constant and the disassociation energy by

$$a = \sqrt{\frac{k}{2A}}. \quad (5.59)$$

(b) The ratio of the cubic to quadratic term in Equation 5.57 is

$$\frac{-Aa^3(r-r_0)^3}{Aa^2(r-r_0)^2} = -a(r-r_0). \quad (5.60)$$

We want the product $a \times (r - r_0)$ to be much less than 1 so that the series in Equation 5.57 converges rapidly and we have confidence applying perturbation methods. A typical value of the parameter a in the Morse potential to describe atom interactions is $a = 27\,nm^{-1}$, and we expect $(r - r_0)$ to be "small" in some sense, because the iron atoms do not stray too far apart. This is encouraging, but to test the relative size of the expansion terms more explicitly, it is convenient to define the dimensionless variable

$$y = a(r - r_0)\sqrt{\frac{A}{\Delta E}}, \tag{5.61}$$

where ΔE is the energy difference defined in Equation 5.53. We can then express the expansion of the Morse potential energy function in terms of y:

$$U_M(y) = -A + \Delta E y^2 - \Delta E \sqrt{\frac{\Delta E}{A}} y^3 + \dots. \tag{5.62}$$

Equation 5.62 is a version of Equation 5.57 that has been recast into "natural units." The ratio of the coefficients of the cubic and quadratic terms is $-\sqrt{\Delta E / A}$, which is a dimensionless function of the two energies that naturally occur in this problem. $\sqrt{\Delta E / A}$ is analogous to the expansion parameter ε in Equation 5.44.

From Equation 5.53, ΔE is the gap between energy levels in the harmonic oscillator, and A is the disassociation energy. We can observe infrared radiation from solid iron, so we can infer that the interatomic "springs" do not break apart whenever the system is excited. This means that $\Delta E \ll A$, and we expect the series in Equation 5.62 to converge rapidly.

To go further, we need to return to experimental data. Representative measured values of $\Delta E = 1.9 \times 10^{-20}\,J$ (which equals to $0.12\,eV$) and $A = 8 \times 10^{-19}\,J$ (or $5\,eV$) yield the dimensionless ratio $\sqrt{\Delta E / A} = 0.155$. This factor is considerably smaller than 1, so the use of perturbation methods is justified for this problem. The higher-order terms in Equation 5.62 can be expressed in terms of increasing powers of $\varepsilon = \sqrt{\Delta E / A}$.

Figure 5.7 suggests that the Morse potential energy function has the qualitative features necessary to model thermal expansion. Unlike the symmetric shape of the parabolic potential for the harmonic oscillator, where the average value of $r - r_0$ is zero independent of the energy level, the shape of the Morse potential is skewed. For higher energy levels, we expect from Fig. 5.7 that the average value of $r - r_0$ increases. Including a cubic perturbation to the potential energy function produces a model that is consistent with an increase in the interatomic distance of the iron atoms with temperature.

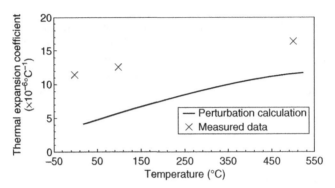

FIG. 5.8. Results for a perturbation calculation of the thermal expansion coefficient for α of iron.

We will outline the perturbation calculation, give its results, but omit the mathematical details. As heat is applied to the bar of iron, its temperature T rises. According to Boltzmann's law, the relative probability of the system to be in the nth quantum state is proportional to

$$P(N) \propto e^{-\frac{E_n}{kT}}, \tag{5.63}$$

where E_n is the energy level of the state and $k = 1.38 \times 10^{-23}$ J/K (not to be confused with the spring constant k). As the temperature increases, the relative probability for the higher energy states increases, as does the expected value of $r - r_0$. This allows us to calculate the thermal expansion coefficient α. Using quantum mechanical perturbation theory, we can calculate the average value of $r - r_0$ for the first few quantum states (as arranged by their energy), starting with the lowest energy state.

Figure 5.8 shows a comparison of the results of the perturbation calculation and the experimental measurements for the thermal expansion coefficient α. The calculated values of α and its general temperature dependence are in qualitative agreement with the observed values for iron. But why is there a factor of 2 discrepancy? Is it because we only kept the cubic term in the expansion of the Morse potential energy function? That is probably not the main source of error; we saw from Equation 5.62 that the series expansion should converge rapidly. A much larger source of error could be our oversimplified model of the lattice of iron atoms. In a real lattice, each atom interacts with multiple neighbors along the three dimensions. Still, obtaining the correct order of magnitude for the thermal expansion coefficient α and predicting how α itself depends on temperature illustrate the value of perturbation methods.

5.6 ISOLATING IMPORTANT VARIABLES

Not all variables are equally important. Some of the variables in an equation (or a set of equations) might be more closely related to the measurable quantities of interest. Other variables might represent quantities that are difficult to interpret or model, and in some cases, those variables can be treated by approximation, or even ignored. When solving such equations, a number of approximation techniques have been developed to exploit the different status among variables. An instructive and well-known example is the analysis of Brownian motion (also called Brownian movement). The treatment of this problem has many analogs that have applications to diverse fields including physics, chemistry, biology, electronics, and economics. The discussion here will be descriptive to illustrate the gist of the approximation techniques used, without getting bogged down in the details of the mathematical analysis, which are fairly complicated. Because many steps are omitted, some of the results are stated rather than derived. More details can be found in Coffey et al. (2004; our discussion of the Langevin equation, Brownian motion, and the fluctuation–dissipation theorem is descriptive and qualitative, and readers who are interested in delving into the mathematical details are referred to this book).

In the nineteenth century, the English botanist Robert Brown viewed small particles suspended in water within a grain of pollen under a microscope. In 1828, he reported a remarkable observation that the particles undergo continual, irregular motion. At first, it was thought that the motion might be related to the living nature of the pollen, but careful experiments by Brown ruled out this possibility. This type of motion, which became known as "Brownian motion," is also displayed by the tiny fat globules that are suspended in milk, or by suspensions of small particles like talcum powder.

Later, in 1905, Albert Einstein provided a clear explanation for the cause of Brownian motion. His explanation is based on the idea that the suspended particle is buffeted about by many water molecules, each of which undergoes random, thermal motion.

A complete mathematical description of Brownian motion would require constructing equations for the suspended particle and for *each* of the many water molecules with which it interacts. This is a hopeless task because the water molecules are too numerous. The great discrepancy between the size, mass, and speed of the suspended particle and the water molecules, however, can be used to simplify the problem with an approximation method.

The individual water molecules that buffet the suspended particle cannot be seen directly with an optical microscope, but Einstein's analysis provided an indirect method to confirm their existence and motion. The variables describing the position and velocity of the suspended particle are important to analyze its motion, but the variables associated with each of the water molecules with which it collides are not. Instead, the approximation methods model the *cumulative* effect of the interaction of the water molecules with the suspended particle.

The following physical assumptions can be built into the approximation. First, the direction of the velocity of the molecules and the location of their collisions with the heavier particle are random. Second, the average time interval between collisions of the water molecules and the suspended particle is much shorter than the time it takes the heavier particle to move an observable distance, even when viewed under a microscope. Finally, the collisions with the heavier particle have a negligible effect on the average speed of the collection water molecules. Instead, their average speed is determined by their temperature T. All three assumptions are valid in practice.

A further simplifying approximation is to use classical mechanics to analyze the problem, despite the tiny size of the molecules. Whenever a water molecule bounces off the suspended particle, the molecule experiences a change in momentum. According to Newton's laws of motion, this in turn imparts a force on the suspended particle. The net result of these countless submicroscopic collisions is to produce a force on the suspended particle that rapidly fluctuates over time. We concentrate on this net fluctuating force and its effect on the variables of the suspended particle while ignoring the variables associated with each of the individual water molecules.

Applying Newton's well-known second law of motion $F = ma$ and the simplifying assumptions to this problem allows us to construct a Langevin equation, named after Paul Langevin, a French physicist who worked in the early part of the twentieth century:

$$m\frac{dv}{dt} = F_C(t) - \frac{v}{\mu} + f_R(t). \tag{5.64}$$

The term on the left-hand side of Equation 5.64 represents mass times acceleration, where m is the mass of the suspended particle, and v is its velocity. The three terms on the right-hand side represent a decomposition of the total force on the suspended particle. The term $F_C(t)$ represents all the "conservative" forces acting on the suspended particle, i.e., not dissipative forces like friction. The conservative forces could include gravity or electrostatic attraction or repulsion. The second term $(-v/\mu)$ models the viscous force on the particle as it moves through the water. The parameter μ is sometimes called the mobility coefficient. The last term on the right-hand side $f_R(t)$ is a rapidly fluctuating force, which is assumed to have a zero time average, and represents the random buffeting by the water molecules.

Various methods, both exact and approximate, have been developed for solving Langevin equations. In some cases, however, useful information can be obtained even without solving the equation directly. Such approaches lead to a powerful relationship known as the *fluctuation–dissipation theorem*. This theorem relates properties describing the variation of the macroscopic variables and was proved by the American physicists Herbert B. Callen and Theodore A. Welton in the 1950s.

Applied to Brownian motion, the fluctuation–dissipation theorem relates the dissipative forces (represented by $1/\mu$) on the suspended particle to the particle's diffusion, i.e., how far it travels in a given time interval due to its random motion. The physical basis of this link is the conservation of energy, which applies regardless of whether or not we track each individual water molecule. The energy lost to friction must be balanced by the energy gained from the random collisions.

In its general form, the fluctuation–dissipation theorem and its proof are rather complicated, and the details are omitted here. One example of a fluctuation–dissipation relationship was derived in Chapter 2 (see Equation 2.58). In the general proof, terms like $f_R(t)$ in Equation 5.64 are assumed to be small enough so that the system responds linearly. Applying the fluctuation–dissipation theorem to the Brownian motion yields a relationship between two measurable, macroscopic quantities for the buffeted particle: the diffusion constant D (in units of meters squared per second) and the mobility coefficient μ (in units of second per kilogram),

$$D = \mu k T, \tag{5.65}$$

where k is Boltzmann's constant and T is the absolute temperature in kelvins. Note in Equation 5.65 that kT has units of mass times velocity squared, or energy, which is expected because it is a measure of thermal energy.

The diffusion coefficient D in Equation 5.65 can be determined, for example, by measuring the root-mean-square distance r_{rms} that the particle covers along one dimension over a time t. Einstein's analysis showed that

$$r_{rms} = \sqrt{2Dt}. \tag{5.66}$$

From Equation 5.64, we can also determine μ by applying a known, constant force $F_C(t) = F_0$. The measured "terminal" average drift velocity v_{term} provides a determination of the mobility coefficient by

$$\mu = \frac{v_{term}}{F_0}. \tag{5.67}$$

In 1908, the French physicist Jean Baptiste Perrin was able to use observations along with the relationships such as Equations 5.65–5.67 to measure the value of Boltzmann's constant k. Then, using the ideal gas law (see Eq. 3.47), Perrin was able to deduce a value of Avogadro's number N_A (i.e., the number of elementary entities in a mole of substance) to within approximately 13% of its true value. His deduction is remarkable considering the astronomical value $N_A \approx 6.022 \times 10^{23}$, which is the number of atoms in 0.012 kg or 12 g of carbon 12. This achievement helped to earn Perrin the 1926 Nobel Prize in physics.

A Langevin equation such as Equation 5.64 serves as the prototype for many related expressions that model situations where many tiny influences accumulate to produce an observable effect on a macroscopic system. Another well-known example of a fluctuation–dissipation relationship arises in electrical engineering. Collisions between the electrons that carry electrical current and the atoms of the metal (e.g., copper) in the wire give rise to a fluctuating voltage $V_R(t)$ known as the Johnson–Nyquist noise, named after John B. Johnson and Harry Nyquist, who were both Swedish-born, American engineers who published their work in 1928. We can write an equation for the voltage for a circuit with inductance L, resistance R, and current $I(t)$:

$$L\frac{dI}{dt} = -RI + V_R(t). \tag{5.68}$$

Note the similarity in the structure of Equations 5.64 and 5.68.

In the case of electronic noise in a circuit, it can be shown that applying the fluctuation–dissipation theorem (in the low-frequency approximation, i.e., $hf \ll kT$) gives the Johnson–Nyquist result for the root-mean-square of the fluctuating voltage

$$V_{\text{rms}} = \sqrt{4RkT\Delta f}, \tag{5.69}$$

where Δf is the frequency range (i.e., the bandwidth measured in hertz) over which we measure the voltage. Equation 5.69 states that even without a voltage source, a small amount of the electronic noise will be present in a circuit. The noise can be reduced by cooling the circuit, which is why some extremely sensitive electronic components are submersed into liquid nitrogen ($T = 77\,\text{K}$, or $-196\,°\text{C}$) to reduce electronic noise. Notice that both Equations 5.65 and 5.69 involve the absolute temperature T of the systems involved. This gives the clue about the involvement of submicroscopic motion, which is characterized by the temperature.

In recent years, models closely related to the Langevin equation have been constructed for fields of study outside of engineering and the physical sciences. The application of this type of mathematical modeling to the fields of economics and finance has had a particularly profound effect. The Black–Sholes model, named after the twentieth-century American economists Fischer Black and his Canadian–American contemporary Myron Scholes, is often written in differential form:

$$dS = \lambda S dt + \sigma S dW. \tag{5.70}$$

In Equation 5.70, S represents the price of a share of stock, t is time, λ is a constant representing the growth rate (assuming $\lambda > 0$), dW is a random increment of "white noise," and σ is a constant representing volatility. Equation 5.70 models the buffeting of share and option prices in the stock market by many complex economic influences, in analogy to the suspended particle undergoing Brownian motion. Analogous to the Langevin equation, the Black–Sholes model contains a term proportional to the share price itself and a rapidly fluctuating term (see exercise 5.12). Scholes was awarded the 1997 Nobel Prize in economics for his work.

In recent decades, talented mathematicians and physicists have been recruited by brokerage houses and hedge funds to perform mathematical modeling of financial markets. These practitioners are sometimes called the "quants," and their field of work has been called computational finance, financial engineering, or "econophysics." The net result of their effort is still being debated. As described in Urstadt (2007; a descriptive account of the profound effects of applying the Langevin equation and related methods to the field of finance is provided in this article), some assign blame for recent problems experienced by the investment industry in the United States and globally linked economies to this reliance on mathematical modeling. Although one important aim of applying the mathematical models to financial markets was to reduce risk, many argue that it inadvertently had the exact opposite effect. Others argue that this application remains valid and useful, but the equations need to be refined to better account for real-world economic situations, including the occurrence of unusual trading patterns that were neglected in the early models.

Topics that arise from the study of Brownian motion provide good examples of how the detailed variation of some variables can be approximated in a mathematical analysis. The same general idea is used in many other applications to make intractable problems manageable. An example from nuclear physics is the scattering of a proton from an oxygen nucleus.

One of the variables we will want to retain in the analysis is the vector representing the separation distance between the incident proton and the center of mass of the oxygen nucleus. If we start considering other variables, however, the problem quickly becomes very daunting. The target oxygen nucleus itself is a complicated object composed of eight protons and eight neutrons, all interacting with each other as well as the incoming proton. The complexity of the 17-particle problem is enormous.

The proton might scatter elastically off of the oxygen nucleus. In that case, the internal structure of the oxygen nucleus remains unchanged. A more complicated situation arises when the incident proton interacts with the oxygen nucleus and forms an excited nuclear state. To fully analyze this problem, the internal degrees of freedom of the 17-particle conglomerate need to be considered.

A considerable simplification can be obtained with an approximation called the nuclear optical model. This model gets its name from an analogy with the behavior of light. Light, when passing though a medium, can be refracted or absorbed. This process can be mathematically modeled with a complex index of refraction $\tilde{n} = n + i\kappa$. The real part n of the index describes refraction, and the imaginary part κ describes absorption.

In the nuclear optical model, an analogy is established between this behavior of light and nuclear scattering. The interaction of the incident proton and target oxygen nucleus is described with a potential energy function that is complex, $U = U_r + iU_i$. The real part U_r governs elastic scattering where the internal composition of the oxygen is not disturbed. The imaginary part U_i describes processes where the incident proton excites the oxygen nucleus to a state with higher energy than its initial or ground state. Although with light, a photon can actually disappear, the proton does not. The "loss" of the proton modeled by the imaginary part of the potential energy function is a simplified representation of the local excitation and detailed involvement of the many internal degrees of the oxygen nucleus. The nuclear optical model, however, is accurate enough to provide useful predictions for many scattering experiments where the final state of the target is observed to be the same as its initial state.

EXERCISES

(5.1) Estimate the following:
   (a) The number of roofing companies in your town. If you live in a rural area, make the estimate for a nearby city with a population of at least 50,000. How would the fact that in cold weather climates, the roofers cannot work during the entire year affect your estimate?
   (b) The number of ping-pong balls that are required to fill an Olympic-sized swimming pool ($25 \times 50\,\mathrm{m}$) to a depth of $2\,\mathrm{m}$. Use $38\,\mathrm{mm}$ for the diameter of each ping-pong ball.

(5.2) A tax-free bond yields 6% interest. If the initial value is $10,000, use the rule of 72 to:
   (a) estimate its value after 12, 24, and 36 years;
   (b) repeat the estimate for 14 years, using the argument that 14 years is 2 years past the doubling milestone at 12 years;
   (c) estimate the bond value at 18 years, using the argument that 18 is midway between 12 and 24, and $\sqrt{2} \approx 1.4$;
   (d) estimate the value at 23 years. Compare your estimates to the exact results. Would using powers of 10 for the estimate be as useful?

(5.3) It is reassuring whenever two independent approaches to solve a Fermi question yield answers that are of the same order of magnitude. Estimate the daily volume of water flowing past the mouth of the Mississippi River by:
   (a) assuming that its water flows at a rate of $2\,\mathrm{m/s}$ (~5 mi/h), and the river is 6-m deep (~20 ft) and 1600-m wide (~1 mi) at its mouth;
   (b) alternatively assuming that the Mississippi River watershed is $1600 \times 1600\,\mathrm{km^2}$ (~1000 × 1000 mi²) in area, which on average receives $30\,\mathrm{cm}$ (~12 in) of rain annually.

State your answers in both cubic meters and cubic miles. Check the answers against an accepted value for the river's output of approximately $1.6 \times 10^9 \, m^3/day$, or $0.4 \, mi^3/day$.

(5.4)

(a) Use the Taylor series expansion of Equation 5.23 to find the next term in Stirling's asymptotic series:

$$n! \approx \sqrt{2\pi n} \, n^n e^{-n}\left(1+\frac{1}{12n}+\ldots\right).$$

(b) Show that the first term in this expansion accurately gives $20!$ to two significant figures, while adding the second term provides accuracy to four significant figures.

(5.5) Show that the specific result, Equation 5.38, agrees exactly with the general approximation for evaluating integrals using the method of stationary phase, Equation 5.41. Would you expect the agreement to be exact when integrating other functions?

(5.6)

(a) Verify Equation 5.20 using a numerical integration technique.

(b) Even though e^{-x^2} does not have an antiderivative among the elementary functions, show that the definite integral

$$I = \int_{-\infty}^{\infty} e^{-x^2} dx$$

can be evaluated analytically. Hint: show that in polar coordinates

$$I^2 = \int_0^{2\pi} d\theta \times \int_0^{\infty} e^{-r^2} r \, dr.$$

(5.7) Suppose a certain type of bacteria is observed to increase its population at a rate of 12% per day. Use the rule of 72 and the conversion factor $2^{10} \approx 1000$ to estimate how much volume a $1\text{-}mm^3$ sample will occupy after 1 year.

(5.8) Show that there are approximately $\pi \times 10^7$ seconds per year. (This number, like the rule of 72, is worthwhile memorizing). Calculate the error using this approximation for:

(a) a non-leap year;

(b) a leap year; and

(c) an average year.

(d) If there are, on average, 100 lightning strikes per second over the entire Earth, estimate how many occur per year.

(5.9) Iterative and recursive techniques are another useful class of approximation methods. For example, we can try to solve the equation $x = f(x)$ for x by making an initial guess x_0 and then by iterating $x_n = f(x_{n-1})$. This procedure converges when $|f'(x)| < 1$. This basic iterative method is related to other numerical methods to solve equations, such as Newton's method.

(a) Select "radians" for the angular mode on a scientific calculator and repeatedly press the "cos" key. The displayed value should converge to approximately 0.73908513. Write an equation that has this value as a root.

(b) Solve the equation $\cos(x^3 + 1) - x = 0$ to three significant figures by iteration, starting with $x_0 = 0.5$ rad.

(5.10) In Example 3.2, we found the dependence of the first-order spherical Bessel function $j_1(r)$ for small values of r by neglecting the r^2y term (in comparison to $2y$) in the equation $r^2\dfrac{d^2y}{dr^2} + 2r\dfrac{dy}{dr} + (r^2 - 2)y = 0$.

With appropriate boundary conditions, we then found that the lowest-order term of the solution is $y_0 = r/3$. We can use iterative techniques to find the next order term in r.

(a) Reintroduce the r^2y term as the zero-order approximation $r^2y_0 = r^3/3$. Iteration then yields the inhomogeneous differential equation

$$r^2\frac{d^2y_1}{dr^2} + 2r\frac{dy_1}{dr} + (-2)y_1 = -r^2y_0 = \frac{-r^3}{3}. \tag{5.71}$$

Substitute a trial solution $y_1 = cr^b$ and show that $b = 3$ and $c = -1/30$.

(b) Compare this result to expansion given in Equation 3.24. Note that once the boundary conditions determine the coefficient of y_0, the coefficient c of the next term in the series is automatically determined because Equation 5.71 is inhomogeneous.

A similar iterative procedure can be used to find higher-order terms in the expansion of the Struve function, which is also described in Example 3.2.

(5.11)

(a) Suppose a bond yields 7.2% per year. Use the rule of 72 to estimate how many years it takes for its value to double, and compare an exact answer using Equation 5.3.

(b) Suppose a population of bacteria grows at a rate of 36% per day. Use the rule of 72 to estimate how many days it takes for the population to double, and compare with the exact answer.

(c) Why is the estimate in part (a) more accurate?

(5.12) Consider the similarities between Equations 5.64, 5.68, and 5.70. Specifically, show that:
(a) if the terms $F_C(t)$ and $f_R(t)$ are zero in Equation 5.64, the value of the velocity exponentially decays to zero;
(b) if $V_R(t) = 0$ in Equation 5.68, then the value of the current exponentially decays to zero;
(c) if the volatility parameter $\sigma = 0$ in Equation 5.70, then the share price grows exponentially.

(5.13)
(a) Write the integrand of Equation 5.43 as $e^{S(\alpha,\beta)}$ and show that $S(\alpha, \beta) = \beta E - \alpha A + \ln(Z(\alpha, \beta))$.
(b) Locate the saddle point of S, which we will call (α_0, β_0). Show that at the saddle point, the following relationships hold:

$$A = \frac{\partial(\ln Z(\alpha, \beta))}{\partial \alpha} \quad \text{and} \quad E = -\frac{\partial \ln(Z(\alpha,\beta))}{\partial \beta}.$$

(These two relationships are precisely the expressions used to obtain the means of the variables A and E in statistical mechanics when the partition function $Z(\alpha,\beta)$ is known. Therefore, the location of the saddle point is associated with these physical quantities.)
(c) Expand the function $S(\alpha,\beta)$ by a Taylor series expansion about the point (α_0,β_0) keeping terms to second order in the differences $(\alpha - \alpha_0)$ and $(\beta - \beta_0)$.
(d) Use the results from Example 7.3b in Chapter 7 to show that the integration over the saddle point gives

$$\rho(A,E) \approx \frac{e^{S(\alpha_0,\beta_0)}}{2\pi\sqrt{\det \mathbf{M}}},$$

where the 2×2 matrix $\mathbf{M}$ is formed from the coefficients in the Taylor series expansion in part (c).
(e) Again use the result of Example 7.3b to show that the explicit form of the 2×2 matrix $\mathbf{M}$ is given by

$$\mathbf{M} = \begin{pmatrix} \dfrac{\partial^2 \ln Z}{\partial \alpha^2} & \dfrac{\partial^2 \ln Z}{\partial \alpha \partial \beta} \\ \dfrac{\partial^2 \ln Z}{\partial \beta \partial \alpha} & \dfrac{\partial^2 \ln Z}{\partial \beta^2} \end{pmatrix}.$$

REFERENCES

Chen G. 2004. The exact solutions of the Schrödinger equation with the Morse potential via Laplace transforms. *Physics Letters A* 326: 55–57.

Coffey WT, Kalmykov YP, and Waldron JT. 2004. *The Langevin Equation: With Applications to Stochastic Problems in Physics, Chemistry and Electrical Engineering*, 2nd ed. Singapore: World Scientific Publishing.

Davydov AS. 1976. *Quantum Mechanics*. Oxford, UK: Pergamon Press.

De Robertis M. 2000. *The Complete Collection of Fermi Questions*. Published online by 4Teachers.ca, Coldwater, Ontario.

Feynman RP, Leighton RB, and Sands M. 2005. *The Feynman Lectures on Physics*. Boston: Addison-Wesley.

Gamelin TW. 2003. *Complex Analysis*. New York: Springer.

Schiff LI. 1968. *Quantum Mechanics*. New York: McGraw-Hill.

Urstadt B. 2007. The blow-up: the quants behind Wall Street's summer of scary numbers. *Technology Review* (published by MIT) 110 (6): 36–42.

Weisskopf VF. 1992. *Knowledge and Wonder*. Cambridge, MA: MIT Press.

FURTHER READING

Benjamin A and Shermer M. 2006. *Secrets of Mental Math: The Mathemagician's Guide to Lightning Calculation and Amazing Math Tricks*. New York: Three Rivers Press.

There are many excellent books describing techniques to aid mental calculations; this book is an example.

Paulos JA. 1988. *Innumeracy: Mathematical Illiteracy and Its Consequences*. New York: Hill and Wang.

A description of the importance of the mathematical literacy that can be developed by posing and solving Fermi questions is given in this book.

Bentley J. 2000. *Programming Pearls*. Reading, MA: Addison-Wesley.

We adapted exercise 5.3 from this book, which has a very informative section on back-of-the-envelope calculations.

Weisskopf VF. 1975. Of atoms, mountains, and stars: a study in qualitative physics. *Science* 187 (4177): 605–612.

A more mathematical, but still very accessible, treatment of estimates based on physical principles is presented in this article.

Horridge A. 2005. The spatial resolutions of the apposition compound eye and its neuro-sensory feature detectors: observation versus theory. *Journal of Insect Physiology* 51: 243–266.

This review article describes the bee's visual system in much more detail.

Zwillinger D. 1993. *Handbook of Integration*. London: Jones and Bartlett.

Integration, including many numerical methods, is discussed in this book.

6

INTRODUCTION TO DIMENSIONAL ANALYSIS AND SCALING

In Chapter 1, we discussed the dimensions of variables and constants that represent physical quantities, and we also discussed several systems of units that are in common use. Methods for performing dimensional checks were presented, and we discussed how those methods can be used to generalize equations and simplify algebraic calculations. In this chapter, we continue that discussion by describing some of the basic aspects of dimensional analysis and scaling.

6.1 DIMENSIONAL ANALYSIS: AN INTRODUCTION

As described in Chapter 1, equations that accurately depict or model the physical world must be dimensionally consistent. This means that all of the terms (i.e., quantities added or subtracted together) in a valid equation must have the same dimensions. Dimensional analysis exploits the same principle: it uses the dimensions of physical quantities to assist in the construction of equations.

Dimensional analysis is based on the concept that the underlying mathematical description of any physical process must be independent of the specific choice of measurement units. Therefore, it should be possible to describe that process in terms of dimensionless quantities. For example, the *motion* of a mass on a spring is the same regardless of whether we choose to measure it in units

Thinking About Equations: A Practical Guide for Developing Mathematical Intuition in the Physical Sciences and Engineering, by Matt A. Bernstein and William A. Friedman
Copyright © 2009 John Wiley & Sons, Inc.

of meters or inches. True, the numerical values of the constants appearing in the equations of motion will differ, but the description of the physical process must be independent of our choice of units.

Through the use of dimensional analysis, it is often possible to develop equations that solve problems that otherwise might be very difficult to attack. The examples presented in this chapter illustrate how dimensional analysis can provide such equations or symbolic expressions up to an unspecified function, or in some cases up to an unspecified constant.

Dimensional analysis can guide the design of experiments, which provide empirical data that can pin down these unspecified functions (or constants), and thus provide the complete solution. The results of the dimensional analysis can also be helpful in guiding one along a theoretical approach to, or numerical analysis of, a problem.

In this section, we present a few simple situations where dimensional reasoning can provide the general insight that is helpful for solving a problem. Then, in the next section, we describe a more systematic procedure for applying dimensional analysis.

Consider the equation of motion for a mass m on a spring with constant k. Newton's second law $F = ma$ yields a second-order differential equation for the displacement x of the mass from its equilibrium position:

$$m\frac{d^2x}{dt^2} = -kx, \tag{6.1}$$

where t represents time. In order for Equation 6.1 to be dimensionally consistent, the ratio k/m must have the dimensions of time^{-2} to match the dimensions of the second-order derivative with respect to time. Therefore, we might anticipate that the frequency f of oscillation (measured in hertz or s^{-1}) is proportional to $\sqrt{k/m}$. This is borne out by the well-known analytic solution of Equation 6.1, which yields the frequency

$$f = \frac{1}{2\pi}\sqrt{\frac{k}{m}} \approx 0.159\sqrt{\frac{k}{m}}. \tag{6.2}$$

There are two physical features coming into play: the influence of the spring's restoring force and the role of inertia. The spring constant k characterizes the former, and from Equation 6.1, we find that it has the dimensions of force per unit length. The mass m characterizes the inertial aspects of the system. Equation 6.2 shows how k and m determine the frequency of oscillation. We can look for analogous quantities in other problems that involve oscillation.

Consider a membrane with mass per unit area η stretched over a circular frame of diameter D with a radial tension ϕ (i.e., force per unit circumferential length). The stretched membrane forms a drumhead, as on a bongo drum or timpani. In analogy to Equation 6.2, we might expect the natural frequencies of the drum to be proportional to $\sqrt{\phi/\eta}$, i.e., the square root of the ratio of a force-related quantity to an inertial quantity. That expression, however, does not have the correct dimensions because η is a mass *per unit area*. The expression can be modified by introducing a physical parameter that has the dimensions of length^2. The diameter squared D^2 of the circular drumhead serves this purpose, and the quantity ηD^2 has the dimension of mass. This simple line of reasoning based on dimensions is again borne out by a more complete analysis. The details are omitted here, but that analysis shows the fundamental frequency of the drumhead's oscillation to be

$$f_0 = 0.765\sqrt{\frac{\phi}{\eta D^2}}. \tag{6.3}$$

Like the spring constant k, the quantity ϕ has dimensions of force per unit length and, like m, the product ηD^2 has dimensions of mass.

We can apply similar dimensional reasoning to the oscillation of a violin string characterized by tension force T and length L. The partial differential equation describing the propagation of a wave on the string is

$$\mu\frac{\partial^2 y}{\partial t^2} + T\frac{\partial^2 y}{\partial x^2} = 0, \tag{6.4}$$

where μ is the mass per unit length of the string, y is the displacement perpendicular to the string, x is the spatial coordinate along the length of the string, and again t represents time. In order for Equation 6.4 to be dimensionally consistent, the dimensions of T/μ must be (length/time)^2, i.e., the dimensions of velocity squared. Therefore, the quantity $(\sqrt{T/\mu})/L$ has the correct dimensions to be a frequency. This simple dimensional reasoning is again on the correct track. Assuming appropriate boundary conditions, the solution of Equation 6.4 demonstrates that the fundamental frequency for the vibration of the violin string is

$$f_0 = \frac{1}{2L}\sqrt{\frac{T}{\mu}}. \tag{6.5}$$

The dimensional analogy between Equation 6.5 and Equation 6.2 is strengthened by observing that the total mass of the violin string is μL and the ratio T/L has the same dimensions as the spring constant k (i.e., force per unit length). We can then recast Equation 6.5 into the form

$$f_0 = \frac{1}{2}\sqrt{\frac{(T/L)}{(\mu L)}}. \tag{6.6}$$

These examples illustrate how we can use dimensional analysis to infer the oscillation frequency in a variety of situations. In each example, we formed a quantity that has dimensions of time^{-1}, or equivalently velocity/length. In each case, the basic structure of the expression for the frequency turned out to be consistent with the results of a more detailed mathematical analysis. Note that dimensional analysis did *not* provide the values of the various proportionality constants. It is much easier, however, to design an experiment to determine the value of the unspecified proportionality constant (e.g., the "0.765" in Eq. 6.3) after the basic analytic structure of the expression has been correctly determined.

Historically, looking for analogous behavior among equations that describe physical systems has led to surprising results. The nineteenth-century Scottish physicist and mathematician James Clerk Maxwell formulated a set of equations to describe electricity and magnetism, building upon the work of the English physicist and chemist Michael Faraday and others. Combining two of Maxwell's equations yields an expression for the *x*-component of the electric field E_x in free space (i.e., a vacuum):

$$\frac{\partial^2 E_x}{\partial z^2} + \varepsilon_0 \mu_0 \frac{\partial^2 E_x}{\partial t^2} = 0. \tag{6.7}$$

Similarly, the *y*-component of the magnetic field B_y satisfies the equation

$$\frac{\partial^2 B_y}{\partial z^2} + \varepsilon_0 \mu_0 \frac{\partial^2 B_y}{\partial t^2} = 0. \tag{6.8}$$

The fundamental constants $\varepsilon_0 = 8.854 \times 10^{-9}$ F/m and $\mu_0 = 1.257 \times 10^{-6}$ H/m are the electrical permittivity and magnetic permeability in a vacuum, respectively. Although Equations 6.7 and 6.8 describe an effect that is very different from the mechanical waves on a violin string, the structure of those equations is strikingly similar to Equation 6.4. From Equation 6.7, the quantity $(\varepsilon_0\mu_0)^{-1}$ must have the dimensions of velocity squared. In fact, Maxwell discovered that the fundamental constants from electricity and magnetism combine to yield $\sqrt{1/\varepsilon_0\mu_0} = 2.998 \times 10^8$ m/s, which is the value of the speed of light c in a vacuum (see exercise 6.1). By working with equations, Maxwell had discovered the electromagnetic wave nature of light! Historical accounts (D'Agostino 2001) indicate that Maxwell used dimensional reasoning to help develop his theory.

6.2 DIMENSIONAL ANALYSIS: A SYSTEMATIC APPROACH

In this section, we present a more systematic way to employ dimensional analysis. Our goal is to use the dimensions of the physical quantities that appear in a problem to obtain information about the functional form of the solution. Some familiarity with basic linear algebra will be helpful to the reader. Fortunately, the systematic application of dimensional analysis is usually quite straightforward, as outlined by the numbered steps listed below and as illustrated by Examples 6.1–6.3.

(1) First, list all the physical quantities associated with the problem. (These are sometimes called the "governing" parameters.) While there is no systematic method to ensure the list is complete, the choice of physical variables, parameters, and constants is often clear from the statement of the problem. At other times, we must rely on our experience, common sense, or intuition to complete the list. As was illustrated in Section 6.1, the physical quantities are often suggested by the variables and constants that appear in a differential equation describing the problem.

(2) If possible, pare down the list generated in step 1 by arguing that some physical quantities have a negligible effect on the physical process and therefore can be ignored.

In many cases, the argument to exclude a particular quantity is made on the basis of scale. For example, a pendulum's length L is a relevant quantity when applying dimensional analysis to find its natural frequency, as described in Example 6.3. Even though the pendulum is made of atoms, the characteristic size of an atom is *not* relevant, because the length of the pendulum is so much greater in scale. This idea is illustrated further in exercise 6.3.

Assume that after reducing the list, there are a total of n relevant physical quantities $q_1, q_2, ..., q_n$ remaining.

(3) Next form all of the independent, *dimensionless* variables from the list q. These dimensionless variables conventionally are denoted by $\pi_1, \pi_2, \pi_3, ..., \pi_k$. Specifically, express each π as a product of the physical variables, by finding a set of exponents $e_1, e_2, e_3, ..., e_n$ such that the product

$$\pi = q_1^{e_1} q_2^{e_2} q_3^{e_3} ..., q_n^{e_n} \tag{6.9}$$

is dimensionless. The solutions for these exponents that make π dimensionless in Equation 6.9 yield a set of coupled, linear equations, which can then be handled with the tools of linear algebra. In particular, the expected number k of dimensionless products can be determined using a method illustrated in Example 6.1.

(4) Finally, apply Buckingham's π-theorem. The π-theorem is named after the American physicist Edgar Buckingham, who published this work in 1914. Luminaries such as Euler, Fourier, Maxwell, Lord Rayleigh, and many others previously made substantial contributions to the subject as well. A proof of the π-theorem is omitted here but is outlined in Bender (1978; a concise treatment of dimensional analysis and scaling, along with many other aspects of mathematical modeling, is provided in this book). The theorem states that a dimensionally consistent function g of the set of physical quantities $q_1, q_2, ..., q_n$,

$$g(q_1, q_2, q_3, \ldots, q_n) = 0, \tag{6.10}$$

can be reexpressed in terms of some other function f of the dimensionless products:

$$f(\pi_1, \pi_2, \pi_3, \ldots, \pi_k) = 0. \tag{6.11}$$

The number k of dimensionless products is equal to the number of physical quantities n minus the number of physical quantities with "independent dimensions." For example, given a subset of the physical quantities $\{q_1, q_2, q_3\}$, the dimensions of q_3 are independent of the other two if we cannot find exponents α and β such that $[q_3] = [q_1^\alpha][q_2^\beta]$. The concept of independent dimensions is quantified later in this section; in Example 6.1, we discuss a simple algorithm to calculate k using standard methods from linear algebra.

To illustrate some of the terminology and definitions that have been introduced in steps 1–4, consider Newton's well-known second law of motion:

$$F = ma. \tag{6.12}$$

There are three physical quantities: $q_1 = F$, $q_2 = m$, and $q_3 = a$. Equation 6.12 is dimensionally consistent because both sides of the equation have the dimensions of force (i.e., measured in units of kg·m·s^{-2} in the International System of Units [SI] system). Equation 6.12 can be cast into the form of Equation 6.10 as

$$g(F, m, a) = F - ma = 0. \tag{6.13}$$

Equation 6.13 can then be reexpressed into dimensionless form as follows:

$$f(\pi_1) = \pi_1 - 1 = 0, \tag{6.14}$$

where $\pi_1 = F/(ma)$. Note that Equation 6.14 is simpler than Equation 6.13 in the sense that it involves only one variable instead of three. Also, the dimensionless ratio π_1 has the same numerical value regardless of whether force is measured in newtons, dynes, pounds, or any other unit of force (provided that mass and acceleration are measured in the corresponding units).

In Equation 6.14, $k = 1$ because there is only a single dimensionless product π_1. In general, k plays a central role in dimensional analysis. For any physical problem, $k \geq 1$; otherwise, we would be unable to construct any dimensionally consistent relationships from the physical quantities q. For example, even the simplest of relationships such as

$$q_1 = q_2 \tag{6.15}$$

implies the existence of the dimensionless product $\pi = q_1 q_2^{-1}$. If we are unable to form at least one dimensionless product π, then we must have missed some physical quantity q and need to go back to step 1 and add to our list.

As explored in exercise 6.4, the number of π's is always less than or equal to the number of q's, so in general, $1 \le k \le n$. Equation 6.11 suggests that dimensional analysis is easier and more useful when k is small, i.e., the fewer the number of dimensionless products, the better. Fortunately, in practice, many problems in dimensional analysis yield $k = 1$ or $k = 2$. When $k = 1$, there is only a single dimensionless product π_1, and it can be shown that the unspecified function f appearing in Equation 6.11 can be expressed in the particularly simple form

$$f(\pi_1) = \pi_1 - \text{const} = 0 \qquad (6.16)$$

or equivalently

$$\pi_1 = \text{const.} \qquad (6.17)$$

If there are two dimensionless products, π_1 and π_2, then Equation 6.11 can be recast into the useful form

$$\pi_1 = h(\pi_2), \qquad (6.18)$$

where h is some unspecified function. Again, we omit the proofs of Equations 6.17 and 6.18, but they are provided in Szirtes (1997; a very systematic treatment of dimensional analysis and scale modeling, including proofs of many theorems mentioned here, is provided in this book).

EXAMPLE 6.1

A light, circular hoop lies horizontally on a frictionless table and spins about its central axis, which is perpendicular to the table. The hoop has radius R and linear density (i.e., mass/unit length) μ, and spins with angular velocity ω measured in radians per second.

(a) Set up linear equations to determine the dimensionless products π.
(b) Cast the linear equations into matrix form, and comment on how, in general, the methods of linear algebra determine the number and the nature of the solutions. Show that only one dimensionless product can be formed for this particular problem.
(c) Find the single dimensionless product π_1 for this problem, and then use dimensional analysis to find an expression for the tension T in the hoop in terms of the other physical quantities.

ANSWER

(a) Because the hoop is supported on a horizontal, frictionless table, we can conclude that quantities such as the acceleration due to gravity and the coefficient of friction are not physically relevant to this particular problem. There are $n = 4$ remaining physical quantities q_i: T, μ, R, and ω. Recalling that radians are dimensionless and decomposing the dimensions of these four quantities into their basic units in the SI system, we find

$$[T] = \text{kg·m·s}^{-2}(\text{force}),$$

$$[\mu] = \text{kg·m}^{-1}(\text{mass per unit length}),$$

$$[\omega] = \text{s}^{-1}(\text{angular velocity}),$$

$$[R] = \text{m (distance).}$$

(6.19)

Because the π is assumed to be dimensionless, its units are "1," and we have

$$[\pi] = \left[T^{e1} \mu^{e2} \omega^{e3} R^{e4} \right] = 1 = 1\,\text{m}^0 \times 1\,\text{kg}^0 \times 1\,\text{s}^0. \tag{6.20}$$

Combining Equations 6.19 and 6.20, each unit (meters, seconds, and kilograms) yields one linear equation. Equating the exponents of meters, we find

$$e_1 - e_2 + 0e_3 + e_4 = 0, \tag{6.21}$$

while equating the exponents of kilograms yields

$$e_1 + e_2 + 0e_3 + 0e_4 = 0, \tag{6.22}$$

and of seconds gives

$$-2e_1 + 0e_2 - e_3 + 0e_4 = 0. \tag{6.23}$$

These three equations are said to be linearly independent because none of them can be generated from a linear combination of the others.

(b) Equations 6.21–6.23 can be expressed more compactly as the matrix product

$$\begin{bmatrix} 1 & -1 & 0 & 1 \\ 1 & 1 & 0 & 0 \\ -2 & 0 & -1 & 0 \end{bmatrix} \begin{bmatrix} e_1 \\ e_2 \\ e_3 \\ e_4 \end{bmatrix} = \begin{bmatrix} 0 \\ 0 \\ 0 \end{bmatrix} \tag{6.24}$$

or $\mathbf{D}\vec{e} = \vec{0}$. Equation 6.24 is an example of a set of homogeneous linear equations, because the vector on its right-hand side is zero. Consequently, Equation 6.24 always has the zero or "trivial" solution:

$$\begin{bmatrix} e_1 \\ e_2 \\ e_3 \\ e_4 \end{bmatrix} = \begin{bmatrix} 0 \\ 0 \\ 0 \\ 0 \end{bmatrix}. \tag{6.25}$$

The zero solution provides no new information, because it merely restates the known fact that $T^0\mu^0\omega^0R^0 = 1$ is a dimensionless product. So, we seek nonzero solutions to Equation 6.24.

Unlike an inhomogeneous system of linear equations, a solution to a homogeneous system can never be unique. Suppose that we do find a nonzero solution $\vec{e}_0$ in Equation 6.24 corresponding to the dimensionless product π_0. Because Equation 6.24 is homogeneous, any scalar multiple $\alpha\vec{e}_0$ will also be a valid solution. The solutions $\vec{e}_0$ and $\alpha\vec{e}_0$, however, are linearly dependent on each other. From Equation 6.9, the scaled solution $\alpha\vec{e}_0$ corresponds to π_0^α. Raising the dimensionless product π_0 to the power α results in yet another dimensionless product, but it does not provide any new information for the dimensional analysis. We seek to find the *linearly independent* solutions to Equation 6.24.

The matrix $\mathbf{D}$ is called the *dimensional matrix*. The dimensional matrix in Equation 6.24, with its rows labeled by the relevant units and its columns labeled by the physical quantities, is

$$\mathbf{D} = \begin{array}{c} \\ \text{m} \\ \text{kg} \\ \text{s} \end{array} \begin{array}{cccc} T & \mu & \omega & R \\ \begin{bmatrix} 1 & -1 & 0 & 1 \\ 1 & 1 & 0 & 0 \\ -2 & 0 & -1 & 0 \end{bmatrix} \end{array}. \tag{6.26}$$

In the case of Equation 6.24, $\mathbf{D}$ is a 3×4 matrix because there are three dimensional units labeling the rows and four physical quantities labeling the columns. Also, in this particular case, all three rows of $\mathbf{D}$ are linearly independent. In the terminology of linear algebra, we say the *rank** of the matrix $\mathbf{D}$ is 3.

---

*As discussed in texts on linear algebra, the number of linearly independent rows of a matrix is always equal to its number of linearly independent columns. Therefore, we can use the general term "rank" instead of row rank or column rank. The concept of rank is particularly useful because there are many readily available numerical algorithms to determine the rank of a matrix.

The number k of dimensionless products π is equal to the number of nonzero, linearly independent solutions of Equation 6.24 for ë. A useful result from linear algebra is

$$k = n - \text{rank}(\mathbf{D}), \qquad (6.27)$$

or the number of linearly independent, nonzero solutions is the number of columns of the dimensional matrix minus its rank.

For this particular example, we have $4 - 3 = 1$, so there is only a single dimensionless product π_1.

(c) Homogeneous systems like Equation 6.24 can typically be solved with simple variable substitution, although more systematic methods like singular-value decomposition can also be used. Because any solution is only defined up to a scalar multiple α, we can arbitrarily set $e_1 = 1$ in Equation 6.23, from which we can infer that $e_4 = -2$. Using Equations 6.22 and 6.23, we then find

$$\begin{bmatrix} e_1 \\ e_2 \\ e_3 \\ e_4 \end{bmatrix} = \begin{bmatrix} 1 \\ -1 \\ -2 \\ -2 \end{bmatrix}. \qquad (6.28)$$

From Equations 6.20 and 6.28, we obtain the single dimensionless product

$$\pi_1 = \frac{T}{\mu \omega^2 R^2}. \qquad (6.29)$$

From Equation 6.17, the desired result of the dimensional analysis is

$$T = \text{const } \mu \omega^2 R^2. \qquad (6.30)$$

Example 6.1 is simple enough that we probably could have guessed the result, Equation 6.30, based on simple dimensional reasoning, as illustrated in Section 6.1. The more systematic method of forming the dimensional matrix and calculating its rank, however, is instructive, because it is often the most effective method to find the π's in more complicated problems.

Dimensional analysis provides Equation 6.30, which indicates that the tension in the hoop scales linearly with the density of the hoop and as the square of its angular velocity and radius. Like the examples presented in Section 6.1, dimensional analysis does not provide the numerical value of the constant.

Having an expression such as Equation 6.30 available as a starting point can greatly reduce the number of measurements required to arrive at the desired equation. In principle, we could determine the value of the constant in Equation 6.30 from a single measurement of one of the quantities (e.g., T) if the values of the other three are known. If we went to the effort to build an experiment setup, however, we might perform several additional measurements varying parameters R or ω to verify that the structure of Equation 6.30 is indeed correct.

The particular problem of the tension in the rotating hoop can be solved exactly using the methods of classical mechanics, so Example 6.1 is primarily for illustrative purposes. The exact solution using mechanical analysis is based on calculating the centripetal force on an element of the hoop. That analysis shows that the value of the constant in Equation 6.30 is equal to 1.

Unlike Example 6.1, for many other problems, we do not have access to the analytic solution, because it is either difficult to find or nonexistent. In those cases, dimensional analysis is particularly useful, as illustrated by the next example. This problem from fluid mechanics was studied in the nineteenth century by John Strutt, the English physicist better known as Lord Rayleigh.

EXAMPLE 6.2

A liquid characterized by density ρ (mass/volume) and surface tension σ emerges slowly in drops from a small circular pipe of radius R. Use dimensional analysis to estimate the weight of each drop. This treatment is adapted from Section 8.3 of Szirtes (1997).

ANSWER

The physical quantities are ρ, R, σ, the mass M of the drop, and the acceleration due to gravity g. The weight of the drop is given by the product Mg. Surface tension σ is a measure of unbalanced molecular cohesive forces at a liquid's surface. Its dimensions are energy/area, or equivalently force/length. In the SI system, the units of surface tension are $kg \cdot s^{-2}$ (see exercise 6.5). Decomposing the dimensions of these five physical quantities into their basic units in the SI system yields

$$[\rho] = kg \cdot m^{-3} \text{(mass density)},$$

$$[M] = kg \text{ (mass)},$$

$$[R] = m \text{ (distance)}. \tag{6.31}$$

$$[\sigma] = kg \cdot s^{-2} \text{(surface tension)},$$

$$[g] = m \cdot s^{-2} \text{(acceleration due to gravity)}.$$

Using the methods outlined in Example 6.1, the dimensional matrix is

$$
\mathbf{D} = \begin{array}{c} \\ \text{m} \\ \text{kg} \\ \text{s} \end{array}
\begin{array}{ccccc} \rho & M & R & \sigma & g \\ \left[\begin{array}{ccccc} -3 & 0 & 1 & 0 & 1 \\ 1 & 1 & 0 & 1 & 0 \\ 0 & 0 & 0 & -2 & -2 \end{array} \right] \end{array}.
\tag{6.32}
$$

The rank of the matrix $\mathbf{D}$ in Equation 6.32 can be readily determined by entering its elements into any matrix calculator, many of which can be accessed from the Internet. The matrix has five columns and a rank of three, so from Equation 6.27, there are $5 - 3 = 2$ dimensionless products. Two nonzero, linearly independent solutions (see exercise 6.6) to the equation $\mathbf{D}\vec{e} = \vec{0}$ are

$$
\vec{e}_2 = \begin{bmatrix} 1 \\ 0 \\ 2 \\ -1 \\ 1 \end{bmatrix} \text{ and } \vec{e}_1 = \begin{bmatrix} 0 \\ 1 \\ -1 \\ -1 \\ 1 \end{bmatrix},
\tag{6.33}
$$

corresponding to two dimensionless products

$$
\pi_1 = \frac{Mg}{\sigma R}
\tag{6.34}
$$

and

$$
\pi_2 = \frac{\rho R^2 g}{\sigma}.
\tag{6.35}
$$

Using Equation 6.18, a general solution can be written:

$$
\pi_1 = \frac{Mg}{\sigma R} = h(\pi_2)
\tag{6.36}
$$

or

$$
Mg = \sigma R \times h\left(\frac{\rho R^2 g}{\sigma} \right),
\tag{6.37}
$$

where h is some unspecified function. Equation 6.37 is as far as dimensional analysis can advance the solution. Heuristic arguments, theoretical analysis, or experimental measurements are needed to learn more about the function h.

Measurements show that, to fair approximation, the weight of the drop is given by the simple relationship

$$Mg \approx 7.6\sigma R. \tag{6.38}$$

Substituting $R = 1\,\text{mm}$ and the empirical value of the surface tension of water ($\sigma \approx 0.072\,\text{kg/s}^2$), Equation 6.38 predicts the mass of the drop is about $0.056\,\text{g}$. We could decrease the surface tension of water by adding a surfactant (i.e., a wetting agent) such as a small amount of liquid soap.

Equation 6.38 implies that the unspecified function h in Equation 6.37 is approximately equal to the dimensionless constant 7.6, independent of the density of the liquid. Equation 6.37 suggests that if we wanted to refine this approximation, we should analyze our measured data by making a plot of $\pi_1 = (Mg/\sigma R)$ versus $\pi_2 = \rho R^2 g/\sigma$. Such a plot is sometimes called "dimensionally homogeneous" because both axes represent dimensionless products. Any dependence that emerges from the plot provides information about the function h. The use of logarithmic scales helps reveal any power-law dependence.

The next problem is related to the various examples of oscillatory behavior discussed in Section 6.1 and illustrates the important concept of a *dimensionally irrelevant* quantity.

EXAMPLE 6.3

A pendulum consists of a mass M on a light rod of length L. The rod is initially released at an angle θ from vertical as shown in Fig. 6.1. Use dimensional analysis to find an expression for the period of oscillation τ, i.e., the inverse of the frequency of oscillation.

ANSWER

The physical parameters are τ, M, L, θ, and, because gravity provides the restoring force, the acceleration due to gravity g. The angle θ is dimensionless, so we have already found one of the π's, and we need not include θ in the list of q's that is used to determine the remaining π's. The dimensional matrix for the remaining physical quantities is

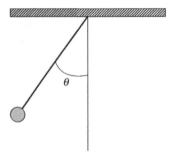

FIG. 6.1. The pendulum analyzed in Example 6.3.

$$\mathbf{D} = \begin{array}{c} \\ \mathrm{m} \\ \mathrm{kg} \\ \mathrm{s} \end{array} \begin{array}{c} L \quad M \quad \tau \quad g \\ \left[\begin{array}{cccc} 1 & 0 & 0 & 1 \\ 0 & 1 & 0 & 0 \\ 0 & 0 & 1 & -2 \end{array}\right] \end{array}. \tag{6.39}$$

Consider the equation associated with the second row of the matrix **D**:

$$0e_1 + e_2 + 0e_3 + 0e_4 = 0. \tag{6.40}$$

Equation 6.40 requires that $e_2 = 0$. That means that in any expression for a π, the factor containing the mass M will be raised to the zeroth power. Because $M^0 = 1$, the parameter M effectively disappears from the expression for any π. The mass M is dimensionally isolated from the other physical quantities as can be seen by the zeros in the second row of the matrix **D** in Equation 6.39, and M is said to be dimensionally irrelevant.

We can remove the quantity M and unit kilograms from consideration to form a reduced dimensional matrix:

$$\mathbf{D'} = \begin{array}{c} \\ \mathrm{m} \\ \mathrm{s} \end{array} \begin{array}{c} L \quad \tau \quad g \\ \left[\begin{array}{ccc} 1 & 0 & 1 \\ 0 & 1 & -2 \end{array}\right] \end{array}. \tag{6.41}$$

The matrix **D'** in Equation 6.39 has three columns and a rank of two, so by Equation 6.27, there is one more π, in addition to the angle θ that was previously identified. We find that

$$\pi_1 = \frac{g\,\tau^2}{L} \tag{6.42}$$

and

$$\pi_2 = \theta. \tag{6.43}$$

From Equation 6.18, dimensional analysis yields

$$\pi_1 = \frac{g\tau^2}{L} = h(\pi_2) = h(\theta). \tag{6.44}$$

The period τ is

$$\tau = \tilde{h}(\theta)\sqrt{\frac{L}{g}}, \tag{6.45}$$

where $\tilde{h}(\theta) = \sqrt{h(\theta)}$. The unspecified function $\tilde{h}(\theta)$ can be determined from a set of measurements of the pendulum's period versus various starting angles. For small oscillations, i.e., when the pendulum is released at a starting angle $\theta \ll 1$ rad, the period of the pendulum is the well-known result

$$\tau = 2\pi\sqrt{\frac{L}{g}}. \tag{6.46}$$

Comparing Equations 6.45 and 6.46, the factor $\tilde{h}(\theta)$ is equal to the constant 2π in this limit. Equation 6.45, which was obtained with dimensional analysis, is valid for greater values of the starting angle as well. In that case, a theoretical analysis shows that $\tilde{h}(\theta)$ is not a constant but instead is a more complicated function of θ that can be expressed in terms of an elliptic integral.

The discovery of dimensionally irrelevant quantities, such as M in Equation 6.39, is a valuable aspect of dimensional analysis. Any such quantities cannot be present in the expression for the π's, so the final expressions cannot depend on them either. That result is illustrated by the absence of M from Equation 6.46. Dimensionally irrelevant quantities further simplify the theoretical analysis and can reduce the number of experiments required to construct an empirical model.

A quantity that is dimensionally irrelevant in one model of a physical situation, however, may no longer be dimensionally irrelevant in a more sophisticated model. In our original consideration of the pendulum, we used a model that neglected the effect of air resistance on its motion. When the effect of air resistance is included, additional physical quantities such as the density ρ of air (in units of kg·m^{-3}) and its viscosity μ (in units of kg·m^{-1}·s^{-1}) need to be considered. When these additional physical quantities are considered, the resulting dimensional matrix **D** will acquire additional columns that have nonzero elements in the row corresponding to the unit "kilograms." Therefore, the solution for the exponent of M is no longer necessarily zero, as the mass M of the pendulum will no longer be dimensionally isolated. We would then expect M to enter into the solution for the pendulum's period τ.

6.3 INTRODUCTION TO SCALING

The term "scaling" is used in several different ways in science and engineering. A common usage refers to the application of multiplicative scale factor, i.e., a scalar multiplier α. When applied to a length, a scale factor stretches it if $\alpha > 1$ or shrinks it if $\alpha < 1$. This usage is related to the idea of scale model, which we discuss first.

Consider the two right triangles M (for "model") and P (for "prototype") depicted in Fig. 6.2. Two objects, such as these triangles, are said to display the property of *physical similarity* if all of the corresponding dimensionless products associated with them are numerically equal. The triangles in Fig. 6.2 are *similar* because the length of each side of the model triangle is scaled by the same factor relative to the corresponding side of the prototype triangle. All of the dimensionless products formed from the lengths appearing in each triangle are numerically equal. In a sense, the scaling can be thought of as expressing the lengths in different units. That change would have no effect on any of the dimensionless products.

The lengths of the hypotenuses of the two similar triangles in Fig. 6.2 differ,

$$H_M \neq H_P, \tag{6.47}$$

but the measure of the corresponding angles (a dimensionless quantity) is numerically equal:

$$\theta_M = \theta_P. \tag{6.48}$$

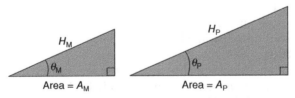

FIG. 6.2. Two similar right triangles provide an example of the property of physical similarity. The corresponding dimensionless products associated with them are numerically equal.

Another dimensionless product can be formed by calculating the ratio of the area A of a triangle to the squared length of its hypotenuse H. As discussed in exercise 6.7, it is straightforward to apply the rules of trigonometry to the triangles in Fig. 6.2 to show that this dimensionless ratio is given by

$$\frac{A}{H^2} = \frac{1}{2} \sin\theta \cos\theta. \qquad (6.49)$$

From Equations 6.48 and 6.49, we can infer that

$$\frac{A_M}{H_M^2} = \frac{A_P}{H_P^2}. \qquad (6.50)$$

Equation 6.50 simply confirms that the corresponding dimensionless product has the same value for the two similar triangles, which should be the case. Interestingly, a demonstration of the Pythagorean theorem can be developed from these ideas, as discussed in exercise 6.8.

The concept of physical similarity is the basis for building and using scale models to solve engineering problems. Often, it is difficult to build a full-sized prototype of a machine or structure, so a scale model is built instead for testing purposes. We usually envision a scale model as being smaller than the engineering prototype, i.e., a miniature version. In some cases, however, an engineering prototype might be difficult to build because it is so *small*, in which case, the scale model could be the larger of the two. The only requirement is that the model and the prototype display the property of physical similarity. We try to design and operate the scale model so that *all* of its dimensionless products π have the same numerical values as in the full-scale prototype. These products are the same π's that were discussed in Section 6.2. Thus, there is a strong link between dimensional analysis and scale modeling. The following example illustrates a typical use of scale modeling and is adapted from Chapter 17 of Szirtes (1997).

EXAMPLE 6.4

Suppose a shipping company wants to estimate how much power P is required to tow an $80 \times 10 \times 6$ m barge with a loaded weight of $W = 2 \times 10^6$ N (approximately 204 metric tons of mass) at a constant speed of $v = 5$ m/s through water. It is too expensive to build a full-scale engineering prototype, so instead, the company intends to build a 40:1 scale-model barge and then take measurements. We say the scale factor for length is $S_L = 1/40$.

The power P required to tow the barge is expended in fighting drag as the barge moves through the water. A preliminary theoretical analysis shows that the relevant physical quantities are P, W, v, the density of water

$\rho \approx 10^3 \, \text{kg/m}^3$, and the acceleration due to gravity $g = 9.8 \, \text{m/s}^2$. The latter two quantities enter into modeling the power because they are important for the formation of the waves and wake produced by the barge, and these, in turn, are the main sources of drag.

(a) What should be the length, width, and height of the scale model? Explain how the weight of the model should be scaled accordingly.
(b) Using the methods described in Section 6.2, form dimensionless parameters π from the physical quantities P, W, ρ, g, and v.
(c) Determine how fast the scale model v_{model} should be pulled through the water. Is the speed of the model 1/40th of the speed of 5 m/s, the speed of the full-scale prototype?
(d) If it is empirically determined that $P_{model} = 0.50 \, \text{W}$ of power is required to pull the scale model through the water, approximately how many watts is required to tow the full-scale barge?

ANSWER

(a) The scale-model barge should be physically similar to a full-scale prototype. Although it need not look exactly like a miniature of the full-scale prototype, ideally, the model is built with the same general shape and material so that the two will behave similarly in the water in terms of drag, stability, etc. Because the scale factor for length has been chosen to be $S_L = 1/40$ along each direction, the size of the scale model is $2.0 \times 0.25 \times 0.15 \, \text{m}$. To keep the density (i.e., mass per unit volume) held constant, the scale factor S_W for the weight of the model must be

$$S_W = S_L^3 \tag{6.51}$$

or $S_W = (1/40)^3 = 1.56 \times 10^{-5}$. The weight of the model is

$$W_{model} = W \times S_W \tag{6.52}$$

or 31.3 N, corresponding to 3.2 kg of mass. These seem to be manageable values, so it is reasonable to proceed. Note that the scale factor for linear dimension (1/40) is *not* numerically equal to the scale factor for weight, because weight is expected to scale as length cubed. This illustrates an important point: while the scale factor S for each physical parameter is a dimensionless number, the numerical values of the various scale factors, in general, differ.

(b) Using the methods described in Section 6.1, we find that the dimensional matrix for this problem is

$$\mathbf{D} = \begin{array}{c} \\ m \\ kg \\ s \end{array} \begin{array}{ccccc} P & W & \rho & v & g \\ \left[\begin{array}{ccccc} 2 & 1 & -3 & 1 & 1 \\ 1 & 1 & 1 & 0 & 0 \\ -3 & -2 & 0 & -1 & -2 \end{array}\right] \end{array}.$$

The matrix has five columns and a rank of 3, so there are two linearly independent, dimensionless products. We can express these as

$$\pi_1 = \frac{Pg^2}{v^7\rho} \tag{6.53}$$

and

$$\pi_2 = \frac{Wg^2}{v^6\rho}. \tag{6.54}$$

(c) Physical similarity between the model and the prototype requires that they operate with the same numerical value of both dimensionless quantities. Consequently, by equating the values of π_2

$$\frac{W_{model}g^2_{model}}{v^6_{model}\rho_{model}} = \frac{Wg^2}{v^6\rho}. \tag{6.55}$$

The acceleration due to gravity and the density of water are physical constants. Consequently, they have the same value for the model and prototype, so their scale factors (e.g., $S_g = g_{model}/g$) are equal to 1. Equation 6.55 therefore reduces to

$$v_{model} = v \times \left(\frac{W_{model}}{W}\right)^{1/6} = v \times S_W^{1/6} = v\sqrt{S_L}. \tag{6.56}$$

Equation 6.56 determines how fast the scale model must be pulled through the water to model the velocity of the full-sized barge. The scale factor for speed is

$$S_v = \frac{v_{model}}{v} = \sqrt{\frac{1}{40}} \approx 0.158. \tag{6.57}$$

Again, note that $S_v \neq S_L \neq S_W$. Because the full-scale prototype will move with speed $v = 5\,\text{m/s}$, Equation 6.57 requires that $v_{model} = 0.79\,\text{m/s}$.

In practice, it might be more convenient to hold the scale-model barge stationary on a tether and have the water stream by it at a steady rate of

0.79 m/s. The power P_{model} required to tow the model could then be calculated by measuring tension T in the tether and multiplying it by water's speed:

$$P_{model} = T v_{model}. \tag{6.58}$$

(d) The requirement of physical similarity must also apply to the other dimensionless product, π_1. From Equation 6.53, we obtain

$$\frac{P_{model} g_{model}^2}{v_{model}^7 \rho_{model}} = \frac{P g^2}{v^7 \rho}. \tag{6.59}$$

From Equation 6.59, the power expected to tow the full-scale prototype is

$$P = P_{model} \left(\frac{v}{v_{model}} \right)^7 = P_{model} \, S_v^{-7} = P_{model} \, S_L^{-7/2}. \tag{6.60}$$

Therefore, the power required to tow the full-scale barge is expected to scale as $(1/40)^{-7/2} = 4.05 \times 10^5$ times the power measured to tow the scale model. So, if we measure that $P_{model} = 0.50\,W$ is required to tow the model, then Equation 6.60 predicts that it requires approximately 0.202 MW, or approximately 271 hp, to tow the full-scale barge.

The *Reynolds number* is a dimensionless parameter that is commonly used to characterize viscous fluid flow. It is named after the English fluid dynamicist Osborne Reynolds, who introduced the concept in the late nineteenth century. The Reynolds number is given by

$$Re_x = \frac{\rho v x}{\mu}, \tag{6.61}$$

where ρ is the density of the fluid (measured in units of $kg \cdot m^{-3}$ in the SI system), v is the mean velocity of the fluid in $m \cdot s^{-1}$, μ is the viscosity of the fluid in $kg \cdot m^{-1} \cdot s^{-1}$, and x is a characteristic length of the physical system measured in meters. Because Re_x is a dimensionless quantity, it has the same numerical value regardless of the choice of the system of units.

The Reynolds number can characterize conditions leading to turbulent flow. For example, in a circular pipe of diameter D, turbulent flow is observed when Re_D is on the order of or greater than 4000. The net value of Re_D determines the onset of turbulence, rather than the specific values of the constituent factors in Equation 6.61. The onset of turbulence is often investigated with scale modeling because of the mathematical complexity of the problem. The values of the Reynolds number (as well as all the other relevant dimensionless parameters) for the model and the engineering prototype should be numerically equal when investigating viscous fluid flow.

A second common use of the word "scaling" refers to the analysis of power-law expressions, i.e., equations of the form

$$y = ax^n. \tag{6.62}$$

For example, in connection with Equation 6.30, we say the tension T *scales* as the square of the radius of the hoop R. Although Equation 6.62 is quite simple, equations like it describe many processes and have useful properties. We begin by discussing the concept of *scale invariance*.

In equations describing certain systems, there is a dimensioned parameter that determines a characteristic measure or *scale* for the problem. The presence or absence of such a parameter has a profound effect on the properties of the system.

Consider two functions $f_1(x)$ and $f_2(x)$, where x has the dimensions of length:

$$f_1(x) = \frac{1}{x^3} \tag{6.63}$$

and

$$f_2(x) = \frac{1}{x^3} e^{-x/L}. \tag{6.64}$$

There is no characteristic length present in f_1. For f_2, however, the parameter L serves that purpose. The properties of $f_2(x)$ depend strongly on how the value of x compares with the characteristic length L. If x is much less than L, the exponential factor is approximately equal to 1, and the function falls off as the inverse cube of x:

$$f_2(x) \approx \frac{1}{x^3}, \quad x \ll L. \tag{6.65}$$

On the other hand, for x much larger than L, the falloff is dominated by the exponential, so the function $f_2(x)$ decays much faster than the inverse cube of x.

The important point is that the function f_2 behaves quite differently depending on the value of the ratio x/L. No such distinction occurs for f_1, which decreases by a factor of 8 every time the distance x doubles, regardless of the starting point. The function f_1 lacks any reference standard for length and is said to be "scale invariant."

Scale invariance is closely related to the property of self-similarity, which is a property displayed by fractals. For example, the snowflake-shaped Koch

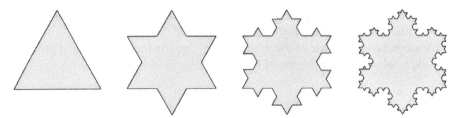

FIG. 6.3. The first four iterations of the Koch snowflake fractal.

curve (Fig. 6.3) looks identical each time one of its sides is magnified by a factor of 3.

We can illustrate scale invariance more quantitatively by considering $f_1(x/b)$, i.e., scaling the argument x by a nonzero parameter b. We find

$$f_1\left(\frac{x}{b}\right) = \frac{1}{(x/b)^3} = b^3\frac{1}{x^3} = b_3 f_1(x). \tag{6.66}$$

Similarly, replacing b with b' yields

$$f_1\left(\frac{x}{b'}\right) = b'^3 f_1(x). \tag{6.67}$$

Eliminating $f_1(x)$ from Equations 6.66 and 6.67 yields

$$f_1\left(\frac{x}{b}\right) = \left(\frac{b}{b'}\right)^3 \times f_1\left(\frac{x}{b'}\right). \tag{6.68}$$

Equation 6.68 states that, with the exception of the multiplicative factor $(b/b')^3$, changing from b to b' makes no difference in the functional form of f_1. Modifying the value of b is the equivalent of expressing x in different units, e.g., millimeters instead of meters. The general power law, Equation 6.62, also displays this property, which explains why it is called *scale* invariant.

On the other hand, the function f_2 is not scale invariant:

$$f_2\left(\frac{x}{b}\right) = \frac{1}{(x/b)^3}e^{-x/(bL)} = b^3 \times \frac{1}{x^3}e^{-x/(bL)} \neq b^3 f_2(x). \tag{6.69}$$

Some physical systems display scale invariance, while others do not. The electrostatic (i.e., Coulomb) force between two charged particles varies as the inverse square of their separation. This is an example of power-law dependence with $n = -2$ in Equation 6.62, so the Coulomb force is scale invariant. This scale invariance has been attributed to the zero mass of the particle that mediates the electrostatic interaction, i.e., the photon. The force between two atomic nuclei, however, does not display scale invariance. The particle that mediates the nuclear interaction is a meson, which serves an analogous role as the photon. The meson* has nonzero mass and provides a characteristic range of 10^{-15} m for the nuclear force. Physicists believe that the massiveness (or lack of mass) of the mediating particle governs the different scaling behavior of the two forces.

Another example of scale invariance in physical systems arises in the study of phase transitions and critical points, which were introduced in Chapter 3. In statistical mechanics, a characteristic distance called the "correlation length" is a measure of how ordered the state of a system is. The correlation length provides a reference scale, similar to L in Equation 6.64, so that the equations describing the system are not scale invariant. As a critical point is approached, however, statistical mechanics predicts that the correlation length goes to infinity. Consequently, the system loses its characteristic length, and the equations describing the system acquire scale invariance. This feature makes the study of systems near their critical points especially interesting. Consider a variety of systems that are characterized by different correlation lengths and behave very differently from each other. As they approach their critical points, however, scale invariance is acquired, and the same scale-invariant equations apply to the various systems. This is an example of the property known as *universality*.

EXAMPLE 6.5

In Chapter 5, the following estimate was derived for the maximum attainable height of a mountain peak on Earth:

$$H \approx \frac{D}{m_{molecule}g}, \tag{6.70}$$

where D is the energy (per molecule) required to deform the rock, $m_{molecule}$ is the molecular mass, and g is the acceleration due to gravity at the surface of the Earth (i.e., $9.8 \, m/s^2$). Using a scaling argument, extend this estimate to some other planet (or one of its moons). Assume that the other planet's mountain is made of the same substance as on Earth, and the surface temperature of the other planet is comparable too.

---

*The word "meson" stems from the word for "middle" in Greek, so named because the meson's mass lies between that of an electron and a proton.

ANSWER

If the other planet's mountain is composed of the same material (e.g., silicon dioxide) as on Earth, its chemistry will be the same, so D will have the same numerical value. The molecular mass $m_{molecule}$ will also be the same. The gravitational acceleration g, however, is expected to differ on the surfaces of the different planets. For Earth, we can calculate the acceleration due to gravity g in terms of the terrestrial mass M_{earth} and radius R_{earth} according to

$$g = \frac{M_{earth}G}{R_{earth}^2}, \tag{6.71}$$

where G is Newton's universal gravitational constant. If the average density (mass/volume) of the Earth is ρ_{earth}, then we can express Equation 6.71 as

$$g = \frac{4\pi G}{3} \rho_{earth} R_{earth}. \tag{6.72}$$

Assuming the other planet is also approximately spherical, we can adapt Equation 6.72 and find its gravitational acceleration:

$$g_{planet} = \frac{4\pi G}{3} \rho_{planet} R_{planet}. \tag{6.73}$$

Combining Equations 6.70, 6.72, and 6.73, the maximum height of the peak on the other planet is

$$H_{planet} = H_{earth} \frac{\rho_{earth} R_{earth}}{\rho_{planet} R_{planet}}. \tag{6.74}$$

Therefore, the maximum attaining height of a mountain on a planet scales inversely with its density and radius. This scaling relationship is a result of the simple power-law dependence for g in Equation 6.70 and ρ and R in Equation 6.73, and the associated scale invariance.

The scaling relationship of Equation 6.74 is not necessarily valid for all values of the parameters. For example, if the model predicts that the height of the mountain is roughly equal to the radius of the planet, the assumption of the planet being spherical implicit in Equation 6.73 is no longer valid. Consequently, R_{planet} is a dimensioned parameter that serves as a characteristic length and can ruin the scale invariance of Equation 6.74, in much the same way the dimensioned parameter L acted in Equation 6.64.

The systematic methods of scale modeling illustrated in Example 6.4 could also be applied to Example 6.5. In Example 6.5, we could interpret the Earth to be the "scale model" for the other planet, which then becomes the "engineering prototype." (In this case, the scale model, Earth, is chosen not because of its size but rather because of its accessibility.) In Example 6.5, the dimensionless parameter that is used to establish physical similarity between the "model" and "prototype" is

$$\pi_1 = \frac{D}{m_{\text{molecule}} H G \rho R}.$$
(6.75)

Because the numerical values of D, m_{molecule}, and G are assumed to be the same on Earth as on the other planet, each of their corresponding scale factors S is equal to 1. Therefore, numerical equality of π_1 on Earth and the planet leads to $H \propto (\rho R)^{-1}$, in agreement with Equation 6.74.

Finally, it is worth noting that the decomposition of the units of physical quantities, e.g., $[\rho] = \text{kg·m}^{-3}$ in Equation 6.31, results in a power-law monomial (i.e., polynomial with a single term) in the fundamental units. This might be obvious, because having a quantity x with polynomial units $[x] = \text{kg} + \text{m}^{-3}$ or transcendental units like $[x] = \sin(\text{kg})$ seems impossible. The power-law monomial property of units can in fact be formally proved (Barenblatt 2003; an excellent explanation of scaling and self-similarity, along with many worked examples from the field of fluid dynamics, is provided in this book), and it establishes a further link between dimensional analysis and scaling. The power-law monomial property was an implicit assumption in the construction of Equation 6.9, which states that each of dimensionless products π is a power-law monomial in the physical quantities q. Therefore, the property of physical similarity leads to power scaling expressions such as Equation 6.60, which in turn display the property of scale invariance. In this way, the topics of dimensional analysis, scale modeling, and scale invariance are closely linked.

The final usage of the word "scaling" that we discuss here is related to the classification of variables as being either *intensive* or *extensive*. As an illustration, we call a multiple-user computer server "scalable" if, by doubling the amount of computer hardware, the system is able to accommodate twice as many users. In this case, the amount of computer hardware *per user* does not change with the number of users and is an example of an "intensive" quantity. Both the number of users and the total amount of computer hardware double, and these are examples of "extensive" quantities.

In Chapter 3, we discussed the equation of state for the ideal gas,

$$PV = NkT,$$
(6.76)

where P is pressure, V is volume, N is the number of gas molecules, T is temperature, and k is Boltzmann's constant. These quantities provide a good illustration of the distinction between intensive and extensive physical variables.

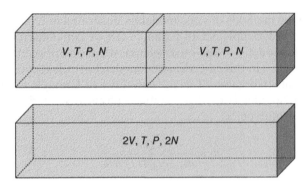

FIG. 6.4. A container of gas is divided into two equal compartments. Each compartment contains N molecules of gas, has volume V, and has temperature T and pressure P. After the central divider is removed, the container has $2N$ molecules and volume $2V$. The value of these *extensive* variables doubles. The gas, however, still has temperature T and pressure P, so the value of those *intensive* variables remains unchanged.

Suppose we have a box of gas divided into two identical regions by a partition, as illustrated in Fig. 6.4. Each side of the box has volume V, pressure P, temperature T, and N gas molecules. If we remove the partition, the combined system has volume $2V$ and $2N$ particles, but it still has temperature T and pressure P. When scaling the amount of material in the system, the quantities V and N are extensive, while T and P are intensive. Note that a ratio of two extensive variables, such as the volume *per* molecule $v = V/N$, does not change, so, like T, it is an intensive quantity.

When constructing equations in thermodynamics, it is often advantageous to use intensive variables, i.e., variables that are independent of the amount of the substance. In principle, such equations apply to systems of any size, even those having an infinite extent. For example, we can recast Equation 6.76 as

$$P(T, v) = \left(\frac{N}{V}\right) kT = \frac{kT}{v}, \qquad (6.77)$$

so that pressure, an intensive quantity, is expressed entirely in terms of intensive variables (also see exercise 3.7 for a discussion about the van der Waals equation of state). Equation 6.77 illustrates a useful rule:

The expression for any *intensive quantity* can always be expressed solely as a function of other intensive variables.

In the case of Equation 6.77, the intensive variable P is expressed in terms of the constant k and the intensive variables T and $v = V/N$. The rule allows us to scale the system (e.g., remove the partition in Fig. 6.4) while ensuring that the value of an intensive variable like pressure P will not change because of a dependence on an extensive variable.

The intensive–extensive property of variables provides information that is independent of dimensional considerations and the other types of scaling discussed in this chapter. For example, dimensionless quantities can be either intensive or extensive. Let G be the Gibbs free energy, as measured in joules. The quantity G is extensive. The product kT is also measured in joules but is intensive. Therefore, the ratio G/kT is an extensive and dimensionless quantity. On the other hand, the ratio $G/(kTV)$ is an intensive quantity that has the dimensions of length^{-3}. The final example of this chapter illustrates the use of intensive and extensive quantities to help construct equations.

EXAMPLE 6.6

A quantity called the "chemical potential" μ that is used in thermodynamics and physical chemistry represents the change in free energy associated with the change in the number of particles (e.g., the number of molecules of gas). The chemical potential is an intensive quantity. In Chapter 3, we mentioned that there are two ways to calculate changes in free energy: at constant T and P, we use the Gibbs free energy G, and at constant T and V, we consider the Helmholtz free energy A. This, in turn, provides two ways to calculate the chemical potential μ. Holding pressure and temperature constant, we have

$$\mu(P, T, N) = \frac{\partial G(P, T, N)}{\partial N}. \tag{6.78}$$

Alternatively, if there is a process where volume and temperature are held constant, we calculate the chemical potential by taking the partial derivative of the Helmholtz free energy:

$$\mu(V, T, N) = \frac{\partial A(V, T, N)}{\partial N}. \tag{6.79}$$

(a) Consider an intensive quantity g representing the Gibbs free energy per particle:

$$g \equiv \frac{G(P, T, N)}{N}. \tag{6.80}$$

Use a general argument about intensive quantities to express the chemical potential in Equation 6.78 in terms of g.

(b) Next, consider the intensive quantity a representing the Helmholtz free energy per particle:

$$a \equiv \frac{A(V,T,N)}{N}. \tag{6.81}$$

Use a general argument about intensive quantities to express the chemical potential in Equation 6.79 in terms of a.

ANSWER

(a) Multiplying G by 1 and using Equation 6.80, we obtain

$$G(P,T,N) = \frac{NG}{N} \equiv N g. \tag{6.82}$$

The pressure P and temperature T are intensive variables, while the Gibbs free energy G and number of particles N are extensive. Because g is the ratio of two extensive quantities, it is intensive. According to the rule, g must be a function of only intensive variables, such as P and T, but specifically *not* N. We will write it as $g(P,T)$. From Equations 6.78 and 6.82, it follows that

$$\mu(P,T) = g(P,T). \tag{6.83}$$

In this way, we obtain an expression for the chemical potential simply by making an argument about intensive variables.

(b) The situation is different for the Helmholtz free energy A, which depends on two extensive variables, N and V. We can still proceed by multiplying by 1:

$$A(V,T,N) = \frac{NA}{N} \equiv Na. \tag{6.84}$$

We know that a is intensive, because it is the ratio of two extensive variables, A and N. Consequently, a can be reexpressed exclusively in terms of intensive variables. Letting $v \equiv V/N$, we can obtain $a(v, T)$. From the chain rule, the partial derivative of some function $f(v)$ with respect to N can be reexpressed in terms of its partial derivative with respect to v:

$$\frac{\partial f}{\partial N} = \frac{\partial v}{\partial N}\frac{\partial f}{\partial v} = -\frac{V}{N^2}\frac{\partial f}{\partial v} = -\frac{v}{N}\frac{\partial f}{\partial v}. \tag{6.85}$$

Combining Equations 6.79, 6.84, and 6.85, it follows that

$$\mu(V, T, N) = a(v, T) - v\frac{\partial a(v, T)}{\partial v}. \tag{6.86}$$

Equation 6.86 provides an expression for the chemical potential in terms of intensive quantities alone.

Free energy and chemical potential might not be very familiar concepts to those who have not studied thermodynamics or physical chemistry. Even so, from a strictly mathematical point of view, Example 6.6 illustrates the utility of exploiting the intensive–extensive property of variables.

EXERCISES

(6.1) Look up the named SI units "farad" and "henry" and express them in terms of the seven fundamental SI units given in Table 1.1. Use dimensional reasoning to form a quantity that has the dimensions of velocity (i.e., with units of meters per second) from the quantities $\varepsilon_0 = 8.854 \times 10^{-9}$ F/m and $\mu_0 = 1.257 \times 10^{-6}$ H/m. Show that the quantity you form is numerically related to $c = 2.99 \times 10^8$ m/s, the speed of light.

(6.2) Consider the physical quantities associated with Newton's second law of motion, i.e., force F, mass m, and acceleration a. The units of these three quantities can be expressed in terms of the SI units kilograms, meters, and seconds.

(a) Construct the dimensional matrix **D** based on these quantities and units.

(b) Show that matrix has only two linearly independent rows, i.e., its rank is 2.

(c) How many dimensionless products can be formed? In other words, what is the value of k?

(d) Explicitly form the dimensionless product(s). What is the value of the undetermined constant that yields Newton's law? Compare your result to Equation 6.14.

(6.3) A powerful explosion releasing energy E creates a shockwave, which then propagates through the air with expanding radius r as a function of time. In 1941, the English physicist Sir Geoffrey Ingram Taylor applied dimensional analysis to this problem. This treatment is adapted from Barenblatt (2003).

Consider the following physical quantities q with dimensions

$[r] = \text{m}$ (radius of shockwave at time t),

$[E] = \text{kg} \cdot \text{m}^2 \cdot \text{s}^{-2}$ (energy of explosion),

$[\rho] = \text{kg} \cdot \text{m}^{-3}$ (density of ambient air),

$[t] = \text{s}$ (time after explosion),

$[P] = \text{kg} \cdot \text{m}^{-1} \cdot \text{s}^{-2}$ (pressure of ambient air),

$[r_0] = \text{m}$ (radius at $t = 0$),

$[\gamma] = 1$ (dimensionless adiabatic index of air ≈ 1.4).

In his analysis, Taylor argued that the initial radius r_0 can be neglected in comparison to r. Also, the ambient air pressure P is expected to be negligible with respect to the air pressure generated by the explosion. So, the quantities P and r_0 were omitted from the dimensional analysis.

(a) Find the dimensionless products π from the remaining physical quantities. The adiabatic index γ is dimensionless, so it is one of the dimensionless variables π. Set up the dimensional matrix for the remaining quantities r, E, ρ, and t.

(b) Label the rows and columns of the dimensional matrix as in Equation 6.26. Using any method of your choice, show that the rank of the dimensional matrix is 3, so there is one additional π besides the adiabatic index γ.

(c) Show that the dimensionless products for this problem can be taken to be $\pi_1 = Et^2/(\rho r^5)$ and $\pi_2 = \gamma$. Then, apply Equation 6.18 to argue that

$$r \propto \left(\frac{Et^2}{\rho} \right)^{1/5}, \tag{6.87}$$

where the proportionality constant is some unspecified function of $\pi_2 = \gamma$. Equation 6.87 was experimentally verified in 1945 by analyzing a series of time-elapsed photographs of an atomic bomb test explosion performed as part of the Manhattan Project. The results showed that the proportionality constant in Equation 6.87 was nearly equal to 1.

(6.4) Suppose that a particular problem has n physically relevant quantities $q_1, q_2, q_3, \ldots, q_n$.
Assume that Equation 6.27 is true. Then $k = n$ implies that the rank of the dimensional matrix must be zero. It is a result from linear algebra that rank($\mathbf{D}$) = 0 only can occur when all of the elements of the dimensional matrix $\mathbf{D}$ are zero. Show that condition, in turn, implies that all of

the q's already happen to be dimensionless products, i.e., $\pi_1 = q_1$, $\pi_2 = q_2, ..., \pi_k = q_n$. (This situation is possible but highly unusual. Typically, at least one of the q's carries dimensions, so that $k \leq n - 1$.)

(6.5) Surface tension σ has dimensions of force per unit length, or energy per unit area. Show that the units of σ in the SI system reduce to $kg \cdot s^{-2}$.

(6.6) The problem stated in Example 6.1 has a generalization in the study of the theory of special relativity, and dimensional analysis can be applied in that case as well. As the linear velocity ωR of an element of the hoop approaches the speed of light c, relativity theory predicts that interesting behavior will be observed. In the relativistic case, the speed of light imposes a natural speed limit and therefore becomes an additional physical quantity q for the problem.

(a) Show that the modified dimensional matrix in Equation 6.26 becomes

$$
\mathbf{D}_{rel} = \begin{array}{c} \\ m \\ kg \\ s \end{array}
\begin{array}{c} T \quad \mu \quad \omega \quad R \quad c \\
\left[\begin{array}{ccccc}
1 & -1 & 0 & 1 & 1 \\
1 & 1 & 0 & 0 & 0 \\
-2 & 0 & -1 & 0 & -1
\end{array} \right]
\end{array}. \tag{6.88}
$$

(b) Calculate the rank of the matrix $\mathbf{D}_{rel}$. Show that the two vectors

$$
\vec{e}_1 = \begin{bmatrix} 1 \\ 0 \\ 2 \\ -1 \\ 1 \end{bmatrix} \text{ and } \vec{e}_2 = \begin{bmatrix} 0 \\ 1 \\ -1 \\ -1 \\ -1 \end{bmatrix}
$$

are solutions to the Equation $\mathbf{D}_{rel}\vec{e} = \vec{0}$.

(c) Show that the two vectors found in part (b) are linearly independent. Hint: if $\vec{e}_1$ and $\vec{e}_2$ were linearly *dependent*, then one of the vectors could be expressed as a scalar multiple of the other: $\vec{e}_1 = \alpha \vec{e}_2$. One way to test this condition numerically is to calculate the ratio of the dot product (i.e., inner product) to the product of the vector magnitudes,

$$
\frac{\vec{e}_1 \cdot \vec{e}_2}{\|\vec{e}_1\| \|\vec{e}_2\|},
$$

and compare the value against 1. The vectors are linearly dependent if the absolute value of the ratio equals to 1.

(d) Show that dimensional analysis provides the expression

$$\frac{T}{\omega^2 \mu R^2} = h\left(\frac{\omega R}{c}\right),\tag{6.89}$$

where h is some unspecified function. Based on the results of Example 6.1, what can you infer about the function h in the limit as $c \to \infty$?

(e) Using the methods of the special theory of relativity, one can show that tension in the hoop is given by

$$T = \frac{\omega^2 \mu R^2}{\sqrt{1 - \left(\frac{\omega R}{c}\right)^2}}$$

Show that this expression is consistent with the results of part (d).

(6.7) Verify Equation 6.49 by calculating one-half the base times height for a right triangle depicted in Fig. 6.2.

(6.8) Consider the right triangle abc depicted in Fig. 6.5a. Consider the physical quantities associated with the triangle: its hypotenuse c, its most acute angle θ, and its area A_c.

(a) Show that there are two dimensionless products associated with those three quantities, and they can be taken to be $\pi_1 = A_c/c^2$ and $\pi_2 = \theta$.

(b) According to Equation 6.18, we have $A_c/c^2 = h(\theta)$, or $A_c = c^2 h(\theta)$, where dimensional analysis does not provide the specific functional form of h. Dropping a perpendicular line, as in Fig. 6.5b, yields two new right triangles with areas A_a and A_b. Argue that we can apply dimensional analysis to those triangles as well to obtain analogous expressions: $A_a = a^2 h(\theta)$ and $A_b = b^2 h(\theta)$. Further, argue that the function $h(\theta)$ has the same numerical value for all three triangles because they are similar. (This last result is verified in exercise 6.7, where $h(\theta)$ is explicitly calculated.)

(c) Use Fig. 6.5b to argue that $A_a + A_b = A_c$. Combine the results of parts (b) and (c) to show that

(a)   (b)

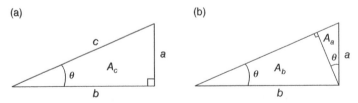

FIG. 6.5. Two right triangles considered in connection with exercise 6.8.

$$c^2 h(\theta) = a^2 h(\theta) + b^2 h(\theta),$$

which implies the Pythagorean theorem, provided that $h(\theta) \neq 0$.

REFERENCES

Barenblatt GI. 2003. *Scaling*. Cambridge, UK: Cambridge University Press.

Bender EA. 1978. *An Introduction to Mathematical Modeling*. New York: John Wiley & Sons.

D'Agostino S. 2001. *A History of the Ideas of Theoretical Physics: Essays on the Nineteenth and Twentieth Century Physics*. Dordrecht, Netherlands: Kluwer Academic Publishers.

Szirtes T. 1997. *Applied Dimensional Analysis and Scaling*. New York: McGraw-Hill.

FURTHER READING

Small CG. 2007. *Functional Equations and How to Solve Them*. Berlin: Springer.

Expressions such as Equation 6.68 are examples of *functional equations*, which arise often in the study of dimensional analysis and scaling. A concise introduction to functional equations is given in this book.

7

GENERALIZING EQUATIONS

We generalize an equation by making changes to it that increase its range of applicability. The resulting generalized equation can then contain multiple special cases embedded within it, so it is usually more useful than the original version. In some ways, the process of generalization is complementary to the processes of approximation and simplification that were discussed in Chapter 5.

Historically, many important advances in mathematics and the physical sciences have resulted from the generalization of an existing equation. Some of these ideas were introduced in Chapter 3, where known, limiting cases were used to guess at the structure of the general solution (see Example 3.1 and the subsequent discussion of blackbody radiation). In this chapter, we illustrate additional methods to generalize equations with a few examples.

Extensions to the binomial formula are discussed in Section 7.1. In Section 7.2, we explore how a simple, concrete example can motivate a more complicated, general expression. In Section 7.3, we discuss recurring themes in generalization, such as increasing the number of dimensions. Section 7.4 deals with the generalization from real to complex variables and illustrates that generalization can sometimes lead to simplification. Finally, in Section 7.5, we discuss the question of *when* to generalize an equation by relating a historical anecdote.

7.1 BINOMIAL EXPRESSIONS

Consider the simple binomial expression for the square of 8:

$$(3+5)^2 = 3^2 + 5^2 + 30 = 64. \tag{7.1}$$

Equation 7.1 is certainly true, but it is not very general and is of limited use. A more general expression is

$$(x+y)^2 = x^2 + y^2 + 2xy, \tag{7.2}$$

where x and y are real variables. Equation 7.2 covers more cases and reduces to Equation 7.1 when $x = 3$ and $y = 5$. We can continue to generalize Equation 7.2 by allowing the exponent to be any positive integer n. This yields the binomial formula, which is a finite series that terminates after $n + 1$ terms:

$$(x+y)^n = y^n + nxy^{n-1} + \frac{n(n-1)}{2!} x^2 y^{n-2} + \ldots + x^n, n = 1, 2, 3\ldots \tag{7.3}$$

When $n = 2$, Equation 7.3 contains three terms and reduces to Equation 7.2.

Continuing to generalize Equation 7.3, consider the case where the exponent is an arbitrary real number r. Provided that $y \neq 0$, let $u = x/y$, yielding

$$(x+y)^r = y^r \left(1+\frac{x}{y}\right)^r = y^r (1+u)^r. \tag{7.4}$$

If we expand $(1 + u)^r$ into an infinite sum using the Taylor series expansion about $u = 0$, we obtain

$$(x+y)^r = y^r \left(1 + ru + \frac{r(r-1)}{2!} u^2 + \frac{r(r-1)(r-2)}{3!} u^3 + \ldots\right). \tag{7.5}$$

To ensure that the series in Equation 7.5 converges, we can choose y to be the element of the pair (x, y) that has the larger absolute value, which implies that $|u| < 1$. Because a generalized equation covers more cases, side conditions, such as $|u| < 1$, frequently need to be specified to ensure its validity. The binomial formula can be generalized further in several different ways. Exercise 7.6 discusses one particular generalization known as the multinomial formula.

The progression from Equation 7.1 to Equation 7.5 illustrates that although more general equations apply to a wider range of situations, they also tend to be more abstract. To better understand the meaning of the general expression, it is sometimes helpful to reduce it to one of the embedded special cases that is more familiar or concrete, e.g., to go from Equation 7.3 to Equation 7.2 by substituting $n = 2$.

Not only do more general expressions tend to be more abstract, but they also tend to be more complicated or messy as well. This again is illustrated by considering the progression from Equation 7.2 to Equation 7.5. Sometimes, the use of compact notation can reduce (or at least hide) the apparent complexity. The standard definition of the binomial coefficient "n choose k" is

$$\binom{n}{k} \equiv {}_nC_k \equiv \frac{n!}{(n-k)!k!},$$
(7.6)

with $0! = 1$, by definition. (Binomial coefficients were used in Example 4.1.) Then, we can rewrite the binomial formula of Equation 7.3 more compactly by using the binomial coefficients defined in Equation 7.3:

$$(x+y)^n = \sum_{k=0}^{n} \binom{n}{k} x^k y^{n-k}, \quad n = 1, 2, 3....$$
(7.7)

We can also apply the same compact notation to the infinite series in Equation 7.5,

$$(x+y)^r = y^r \sum_{k=0}^{\infty} \binom{r}{k} u^k = \sum_{k=0}^{\infty} \binom{r}{k} x^k y^{r-k},$$
(7.8)

where the generalized binomial coefficient

$$\binom{r}{k} = \frac{r \times (r-1) \times (r-2)...(r-k+1)}{k!}$$
(7.9)

is defined for non-integer values of r, and reduces to Equation 7.6 when r is an integer. Note the similarity between Equations 7.7 and 7.8 when the generalized notation for the binomial coefficient is used.

7.2   MOTIVATING A GENERAL EXPRESSION

Sometimes a general expression looks so complicated that the process used to derive it seems mysterious. For example, there are a variety of identities that involve the binomial coefficient, such as

$$\sum_{k=0}^{n} \binom{n}{k} \times \binom{N-n}{n-k+1} = \binom{N}{n+1}, \quad N \geq 2n+1$$
(7.10)

and other related identities described in exercises 7.8 and 7.11. Equation 7.10 looks complicated, especially when all of the binomial coefficients are expanded into factorials with the use of Equation 7.6

$$\sum_{k=0}^{n} \frac{n!(N-n)!}{(k!)(n-k)!(N-2n+k-1)!(n-k+1)!} = \frac{N!}{(N-n-1)!(n+1)!}, \quad N \geq 2n+1.$$

$$(7.11)$$

We can gain confidence that Equation 7.10 is true by numerical substitution of a few sample values of n and N. For example, when $N = 6$ and $n = 2$, it is straightforward to verify that both sides of Equation 7.10 are equal to 20. The purpose of the next example is to illustrate how an expression such as Equation 7.10 can be obtained in a few steps by generalizing a simple, special case.

EXAMPLE 7.1

(a) Consider a standard deck of 52 playing cards with four aces. How many different five-card poker hands are there? How many of these poker hands contain k aces, where $k = 0, 1, 2, 3$, or 4?

(b) Verify that the total number of possible poker hands is equal to the total number of poker hands calculated in (a) containing the various number of aces.

(c) Use the result of (b) to motivate the general identity given in Equation 7.10.

ANSWER

(a) The order of the cards does not matter within a poker hand. Therefore, the binomial coefficient provides a count of the total number of distinct five-card poker hands: $\binom{52}{5} = 2,598,960$.

Because there are four aces in the 52-card deck, there are 48 remaining cards that are not aces. Therefore, the number of five-card poker hands with k aces is $\binom{4}{k} \times \binom{48}{5-k}$. The first factor represents the number of ways to choose the k aces among the four suits, and the second factor represents the number of ways of choosing the remaining $(5 - k)$ cards, none of which are aces.

(b) Using these expressions, we can verify that the sum of the number of poker hands with $0, 1, 2, 3$, or 4 aces is equal to the total number of possible hands. Specifically, we calculate that

$$\binom{4}{0}\times\binom{48}{5}+\binom{4}{1}\times\binom{48}{4}+\binom{4}{2}\times\binom{48}{3}+\binom{4}{3}\times\binom{48}{2}+\binom{4}{4}\times\binom{48}{1}$$

$$= 1,712,304 + 778,320 + 103,776 + 4,512 + 48 = 2,598,960 = \binom{52}{5}. \qquad (7.12)$$

A skilled poker player could use these results to calculate that the probability that five random cards contain a pair of aces is 103,776/2,598,960, or about 0.04 = 4%.

(c) We can rewrite Equation 7.12 in a more compact form:

$$\sum_{k=0}^{4}\binom{4}{k}\times\binom{48}{5-k}=\binom{52}{5}. \qquad (7.13)$$

It no longer requires a big leap to arrive at Equation 7.10. We identify that $N = 52$ is the total number of cards in the deck, $n = 4$ is the number of aces in the deck, and $n + 1 = 5$ is the number of cards per hand. Equation 7.10 then becomes a fairly straightforward generalization of Equation 7.13.

Example 7.1 illustrates how a complicated, general equation sometimes can be obtained from (or at least motivated by) a simple idea. It is a simple concept that the number of aces in any five-card poker hand must be 0, 1, 2, 3, or 4. That observation along with applying the combinatorial (i.e., counting) property of the binomial coefficient yields Equation 7.13. That result, in turn, motivates the more general expressions given by Equations 7.10 and 7.11. The motivation of a more complicated expression by a simpler, concrete example is a common strategy for generalizing an equation.

Sometimes, technical books and journal articles state a general equation, such as Equation 7.10, without much detailed explanation about the thought process used to derive or motivate it. That terse style is often necessary because it conveys a great amount of information while using the least possible space. It is reassuring to keep in mind, however, that an expression like Equation 7.11 rarely, if ever, occurs spontaneously to anyone in its full-blown form, even to experienced scientists, engineers, and mathematicians.

7.3 RECURRING THEMES

Equations can be generalized in many ways, but in practice, some of the same themes seem to recur. The discussion about the binomial expansion illustrates a typical path: the exponent is a specific numerical value in Equation 7.2, then an arbitrary positive integer in Equation 7.3, and finally an arbitrary real number in Equation 7.3. Some other common paths of generalization are from lower to higher dimensions, from fewer to more variables or degrees of freedom, from real to complex numbers, and from a number to a matrix. The examples and exercises in the remainder of the chapter illustrate these themes, with the next example taken from analytic geometry.

It is well known that the area of a triangle is one-half the product of the lengths of its base and height:

$$a = \frac{1}{2} b \times h. \tag{7.14}$$

Sometimes, however, we do not know the values of these particular quantities. Instead, we might be given the lengths of its three sides, in which case, we can calculate the area of the triangle by applying the formula credited to the ancient Greek mathematician Heron of Alexandria. Alternatively, we might be given the Cartesian coordinates for the three vertices of a triangle that lies in the xy-plane. In that case, there is another well-known formula for the area of the triangle, which involves a determinant. Calculating the area of a triangle using this information illustrates another approach to generalizing equations.

Suppose a triangle lies in the xy-plane, and the Cartesian coordinates of its three vertices are (x_1, y_1), (x_2, y_2), and (x_3, y_3). The area a of the triangle is given by the determinant

$$a = \frac{1}{2} \times \left| \det \begin{bmatrix} x_1 & y_1 & 1 \\ x_2 & y_2 & 1 \\ x_3 & y_3 & 1 \end{bmatrix} \right|, \tag{7.15}$$

where the absolute value sign in Equation 7.15 is needed to ensure that $a > 0$, because the area is assumed to be a positive quantity.

First, we attempt to increase the plausibility of Equation 7.15 by considering a few special cases. This process is related to the method of Example 3.1, where several guesses were made about the structure of the general result, and then some were eliminated based on special cases. (More special cases that bolster the plausibility of Eq. 7.15 are discussed in exercise 7.12). Next, the expression for the area of a triangle is generalized to higher dimensions, which yields a related expression for the volume V of a tetrahedron, i.e., a four-sided pyramid. Finally, we consider the extension to dimension higher than three.

EXAMPLE 7.2

(a) Verify that Equation 7.15 gives the correct area a for the triangle depicted in Fig. 7.1: One vertex lies on the origin, and a second

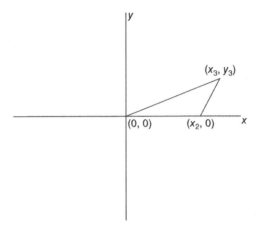

FIG. 7.1. Equation 7.15 gives the area of the triangle lying in the xy-plane. The simple, special case that is illustrated here serves as a check and increases the plausibility that the expression is true in general.

vertex lies on the x-axis. The third vertex can lie anywhere in the xy-plane.

(b) Verify that Equation 7.15 is correct if one vertex lies on the origin, but the other two can lie anywhere in the xy-plane. Figure 7.2 illustrates an example of this configuration.

(c) Consider a lower-dimensional problem. Express the length of line segment on the x-axis as a determinant that is analogous to Equation 7.15.

(d) Based on parts (a–c), generalize Equation 7.15 to higher dimensions. Find an expression for the volume of a tetrahedron given the coordinates of its four vertices. Verify that the expression is correct for the special case illustrated in Fig. 7.3. How could this expression be generalized further to four, and even higher, dimensions?

ANSWER

(a) This special case is illustrated in Fig. 7.1. Substituting the coordinates of the vertices into Equation 7.15 yields

$$a = \frac{1}{2}\det\begin{bmatrix} 0 & 0 & 1 \\ x_2 & 0 & 1 \\ x_3 & y_3 & 1 \end{bmatrix} = \frac{1}{2}|x_2 y_3| \tag{7.16}$$

in agreement with Equation 7.14, i.e., one-half base times height.

Special cases such as Equation 7.16 increase the plausibility that Equation 7.15 is a general result. In order to convince ourselves, however, we

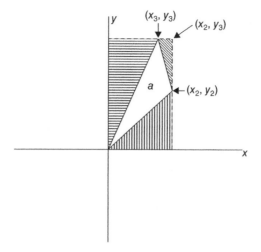

FIG. 7.2. A situation where none of the sides of the triangle is parallel to the x- or y-axis provides a more general test of Equation 7.15.

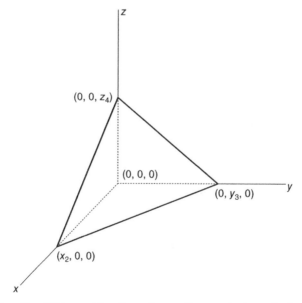

FIG. 7.3. Equation 7.24 provides the volume of a tetrahedron. As in Fig. 7.1, the special case that is illustrated here serves as a check and increases the plausibility that the expression is true in general.

probably have to examine at least one more general case where none of the sides of the triangle lies parallel to the x- or y-axis.

(b) It is helpful to circumscribe a rectangle around the triangle, as illustrated in Fig. 7.2. This rectangle has a total area $x_2 y_3$. It contains the triangle of unknown area a, in addition to three right triangles that are shaded in the figure. Equating the area of the rectangle to the area of the four triangles that comprise it yields

$$x_2 y_3 = a + \frac{1}{2}(x_3 y_3 + x_2 y_2 + (x_2 - x_3)(y_3 - y_2)). \tag{7.17}$$

Solving Equation 7.17 for the desired area, we obtain

$$a = \frac{1}{2}(x_2 y_3 - x_3 y_2). \tag{7.18}$$

Equation 7.18 gives the same result as the determinant

$$a = +\frac{1}{2} \det \begin{bmatrix} 0 & 0 & 1 \\ x_2 & y_2 & 1 \\ x_3 & y_3 & 1 \end{bmatrix} = \frac{1}{2}(x_2 y_3 - x_3 y_2). \tag{7.19}$$

(c) Naturally, the length of line segment that lies between x_1 and x_2 on the x-axis is given by $|x_2 - x_1|$. Using Equation 7.15 as a guide, we can write this same expression in terms of a determinant of a 2×2 matrix,

$$\ell = \left| \det \begin{bmatrix} x_1 & 1 \\ x_2 & 1 \end{bmatrix} \right|, \tag{7.20}$$

where the absolute value operation ensures that the length ℓ is a positive quantity.

(d) If the vertices of the tetrahedron are given by (x_i, y_i, z_i), with $i = 1, 2, 3,$ and 4, then by generalizing Equations 7.20 and 7.15, we might guess that the volume of the tetrahedron is given by

$$V = \left| k \times \det \begin{bmatrix} x_1 & y_1 & z_1 & 1 \\ x_2 & y_2 & z_2 & 1 \\ x_3 & y_3 & z_3 & 1 \\ x_4 & y_4 & z_4 & 1 \end{bmatrix} \right|. \tag{7.21}$$

We can determine the proportionality constant k by recalling the three-dimensional analog of Equation 7.14 for the volume of a tetrahedron,

$$V = \frac{1}{3} A \times h, \tag{7.22}$$

where A is the area of the base and h is the height of the tetrahedron.

Usually, the determinant of a 4×4 matrix is quite messy to evaluate, but the special case illustrated in Fig. 7.3 yields a simplified form:

$$V = \left| k \times \det \begin{bmatrix} 0 & 0 & 0 & 1 \\ x_2 & 0 & 0 & 1 \\ 0 & y_3 & 0 & 1 \\ 0 & 0 & z_4 & 1 \end{bmatrix} \right| = |k \times x_2 \times y_3 \times z_4|. \tag{7.23}$$

The base of the tetrahedron depicted in Fig. 7.3 has area $A = x_2 y_2 / 2$, and its height is $h = z_4$. Applying this result to Equations 7.22 and 7.23, we can infer that

$$V = \frac{1}{6} \left| \det \begin{bmatrix} x_1 & y_1 & z_1 & 1 \\ x_2 & y_2 & z_2 & 1 \\ x_3 & y_3 & z_3 & 1 \\ x_4 & y_4 & z_4 & 1 \end{bmatrix} \right|. \tag{7.24}$$

Note that for a one-dimensional line segment, the coefficient in Equation 7.20 is 1; for a two-dimensional triangle, the coefficient is one-half; and for a three-dimensional tetrahedron, the coefficient is one-sixth. Further generalizing Equations 7.20, 7.15, and 7.24, we could extrapolate that the Euclidean volume of an N-dimensional object with $N + 1$ vertices is given by

$$V_H = \frac{1}{N!} \left| \det \begin{bmatrix} x_1 & y_1 & z_1 & \cdot & q_1 & 1 \\ x_2 & y_2 & z_2 & \cdot & \cdot & 1 \\ \cdot & & & & & 1 \\ \cdot & & & & & 1 \\ \cdot & & & & & 1 \\ \cdot & & & & \cdot & 1 \\ x_{N+1} & y_{N+1} & z_{N+1} & \cdot & q_{N+1} & 1 \end{bmatrix} \right|. \tag{7.25}$$

The Euclidean volume of an object in four or more dimensions is often called its "hyper-volume." For example, the Euclidean hyper-volume of a rectangular prism in four dimensions is the product of the lengths of its four sides.

Example 7.2 illustrates several principles commonly used when generalizing equations. First, as illustrated in parts (a) and (b), simpler, special cases are examined before tackling the complicated, general result. In part (b), one vertex of the triangle was placed at the origin, which simplified the calculation of the determinant in Equation 7.19. The approach of tackling a simpler problem that ultimately demonstrates the generalized result is commonly used in scientific, engineering, and mathematical derivations. Often, the phrase "without loss of generality" appears when the simplifying assumption is stated.

In this particular example, it is very plausible that the area a of the triangle does not depend on the choice of the origin for the coordinate system, so many would agree that the simplified derivation in part (b) does indeed demonstrate the general result. We could say, "Without loss of generality, the first vertex can be placed at the origin." It is never a bad idea, however, to be skeptical and scrutinize these simplifying assumptions carefully. A more general demonstration of Equation 7.15, using properties of determinants and rotation matrices, is outlined in exercise 7.10.

Example 7.2 also illustrates generalization to higher dimension. Although we have not proved the results, Equations 7.24 and 7.25 are in fact correct. (Further arguments are presented in exercise 7.10.) Although these results can yield quite complicated expressions when the determinants are expanded, they are natural extensions of the one- and two-dimensional cases. It is also interesting that while the equations are rather easily generalized to dimensions $N > 3$, it becomes difficult or impossible to depict the higher-dimensional objects with diagrams.

A useful part of the process of generalizing these equations is to "take a step back." In the case of generalization to higher dimension, this meant examining the simple, one-dimensional case in part (c) after validating the two-dimensional result. Probably no one would bother to use a 2×2 determinant to calculate a simple line length, but the structure of Equation 7.20 is quite useful because it gives us confidence that our hunch about the generalization of this problem to higher dimensions is on the right track.

Generalization to higher dimensions, especially using matrices and determinants, is a common procedure. The next example illustrates the application of ideas similar to those used in Example 7.2 to an important problem from calculus. In Chapter 5, we discussed Gaussian definite integrals such as

$$\int_{-\infty}^{\infty} e^{-\frac{ax^2}{2}} dx = \sqrt{\frac{2\pi}{a}}, \tag{7.26}$$

where a is a real, positive constant. This result can be further generalized to multiple dimensions, as illustrated by the next example.

EXAMPLE 7.3*

(a) Show that the two-dimensional integral

$$J = \int_{y=-\infty}^{\infty} \int_{x=-\infty}^{\infty} e^{-\frac{1}{2}\left(ax^2 + by^2\right)} dx \, dy \tag{7.27}$$

*Example 7.3 assumes familiarity with several results from linear algebra.

evaluates to $J = \dfrac{2\pi}{\sqrt{ab}}$, where a and b are both real, positive constants.

(b) Generalize the result in part (a) to show that

$$J_2 = \int\limits_{y=-\infty}^{\infty} \int\limits_{x=-\infty}^{\infty} e^{-\frac{1}{2}\left(ax^2+by^2+2cxy\right)} dx\, dy = \frac{2\pi}{\sqrt{ab-c^2}} = \frac{2\pi}{\sqrt{\det \mathbf{M}}}, \quad (7.28)$$

where $\mathbf{M}$ is a 2×2 real, symmetric matrix,

$$\mathbf{M} = \begin{bmatrix} a & c \\ c & b \end{bmatrix}, \quad (7.29)$$

with $a > 0$ and $\det(\mathbf{M}) = ab - c^2 > 0$. Note: these last two conditions ensure that the eigenvalues of $\mathbf{M}$ are positive. (Eigenvalues were previously discussed in Example 4.3.)

ANSWER

(a) The two-dimensional function $f(x, y) = \exp[-(ax^2 + by^2)/2]$ is *separable* in x and y, that is, it can be written as the product of two independent functions:

$$f(x,y) = e^{-\frac{ax^2}{2}} e^{-\frac{by^2}{2}}. \quad (7.30)$$

Consequently, the integral in Equation 7.27 also separates into the product of two independent integrals, each of which can be evaluated by using Equation 7.26:

$$I = \left(\int\limits_{x=-\infty}^{\infty} e^{-\frac{ax^2}{2}}\, dx \right) \times \left(\int\limits_{y=-\infty}^{\infty} e^{-\frac{by^2}{2}}\, dy \right) = \sqrt{\frac{2\pi}{a}} \times \sqrt{\frac{2\pi}{b}} = \frac{2\pi}{\sqrt{ab}}. \quad (7.31)$$

(b) Because of the cross-term $2cxy$ in the exponential, the integrand is no longer separable, so the method of part (a) no longer works. We can evaluate the integral, however, by applying several well-known results from linear algebra. First, recognize that the exponent in Equation 7.28 contains the *quadratic form*

$$Q(x, y) = ax^2 + by^2 + 2cxy. \quad (7.32)$$

The value of quadratic form Q is a number, but we can also think of it as a 1×1 matrix. It can be expressed as the product of three matrices with dimensions 1×2, 2×2, and 2×1:

$$Q(x, y) = [x \quad y]\begin{bmatrix} a & c \\ c & b \end{bmatrix}\begin{bmatrix} x \\ y \end{bmatrix}. \tag{7.33}$$

Equation 7.33 can be written in the more compact matrix form

$$Q(x, y) = \mathbf{x}^T \mathbf{M} \mathbf{x}, \tag{7.34}$$

where the superscript "T" indicates matrix transpose.

The *principal axes theorem* from linear algebra states that because $\mathbf{M}$ is a real, symmetric matrix, there exists a rotated set of coordinates (u, v) such that the cross-term in the quadratic form drops out:

$$Q(u, v) = \lambda_1 u^2 + \lambda_2 v^2 = [u \quad v]\begin{bmatrix} \lambda_1 & 0 \\ 0 & \lambda_2 \end{bmatrix}\begin{bmatrix} u \\ v \end{bmatrix} = \mathbf{u}^T \Lambda \mathbf{u}, \tag{7.35}$$

where λ_1 and λ_2 are the eigenvalues of $\mathbf{M}$. Equation 7.35 has a simple geometric interpretation, as illustrated by Fig. 7.4. The equation

$$Q(u, v) = \lambda_1 u^2 + \lambda_2 v^2 = 1 \tag{7.36}$$

describes an ellipse centered at the origin, assuming the eigenvalues λ_1 and λ_2 are positive, as stated previously. The length of its semimajor axis is the greater of the pair $(1/\sqrt{\lambda_1}, 1/\sqrt{\lambda_2})$, while the length of its semiminor axis is the smaller. If the two eigenvalues are equal to each other, then Equation 7.36 represents a circle. Using Equations 7.31 and 7.35, and the identity

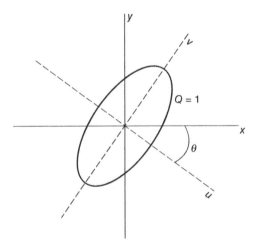

FIG. 7.4. The *uv*-coordinate system is aligned with the principal axes of the ellipse. In this coordinate system, the cross-term in the quadratic form vanishes, as can be seen from Equation 7.35.

$dx\, dy = du\, dv$ (see exercise 7.2), the integral J_2 can be evaluated in the uv-coordinate system, where it is separable:

$$J_2 = \int\limits_{v=-\infty}^{\infty} \int\limits_{u=-\infty}^{\infty} e^{-\frac{1}{2}\left(\lambda_1 u^2 + \lambda_2 v^2\right)} du\, dv = \frac{2\pi}{\sqrt{\lambda_1 \lambda_2}} = \frac{2\pi}{\sqrt{\det \Lambda}}. \quad (7.37)$$

From Fig. 7.4, we can express this rotation of the xy-coordinate system as

$$\begin{bmatrix} u \\ v \end{bmatrix} = \begin{bmatrix} \cos\theta & -\sin\theta \\ \sin\theta & \cos\theta \end{bmatrix} \begin{bmatrix} x \\ y \end{bmatrix}, \text{ or } \mathbf{u} = \mathbf{R}\mathbf{x}, \quad (7.38)$$

where θ is the angle between the u- and x-axes in Fig. 7.4. The rotation matrix $\mathbf{R}$ in Equation 7.38 is orthogonal, i.e., its transpose is also its own inverse:

$$\mathbf{R}^{\mathrm{T}}\mathbf{R} = \mathbf{R}\mathbf{R}^{\mathrm{T}} = \begin{bmatrix} 1 & 0 \\ 0 & 1 \end{bmatrix} = \mathbf{I}, \quad (7.39)$$

where $\mathbf{I}$ is the identity matrix.

Inserting identity matrices from Equation 7.39 into Equation 7.34 yields

$$Q = \mathbf{x}^{\mathrm{T}}\mathbf{M}\mathbf{x} = \mathbf{x}^{\mathrm{T}}\left(\mathbf{R}^{\mathrm{T}}\mathbf{R}\right)\mathbf{M}\left(\mathbf{R}^{\mathrm{T}}\mathbf{R}\right)\mathbf{x}. \quad (7.40)$$

The associative property of matrix multiplication gives

$$Q = \left(\mathbf{x}^{\mathrm{T}}\mathbf{R}^{\mathrm{T}}\right)\left(\mathbf{R}\mathbf{M}\mathbf{R}^{\mathrm{T}}\right)\left(\mathbf{R}\mathbf{x}\right) = \mathbf{u}^{\mathrm{T}}\left(\mathbf{R}\mathbf{M}\mathbf{R}^{\mathrm{T}}\right)\mathbf{u}. \quad (7.41)$$

The numerical value of the quadratic form Q is independent of whether it is evaluated in the (x, y) or (u, v) coordinate systems. Comparing the last term in Equation 7.41 with Equation 7.35, we can infer that the diagonal matrix Λ is obtained from the similarity transformation

$$\Lambda = \mathbf{R}\mathbf{M}\mathbf{R}^{\mathrm{T}}. \quad (7.42)$$

(Eq. 7.42 can be stated as part of the principal axis theorem.) Finally, we can use the property of determinants that

$$\det(\mathbf{A}\mathbf{B}) = \det(\mathbf{A}) \times \det(\mathbf{B}), \quad (7.43)$$

where $\mathbf{A}$ and $\mathbf{B}$ are both $N \times N$ matrices. Equations 7.42 and 7.43 yield

$$\det(\Lambda) = \det\left(\mathbf{R}\mathbf{M}\mathbf{R}^{\mathrm{T}}\right) = \det(\mathbf{R}) \times \det(\mathbf{M}) \times \det\left(\mathbf{R}^{\mathrm{T}}\right). \quad (7.44)$$

From Equation 7.39, the product of determinants reduces to

$$\det(\mathbf{R}) \times \det(\mathbf{R}^{\mathsf{T}}) = \det(\mathbf{R}\mathbf{R}^{\mathsf{T}}) = 1. \tag{7.45}$$

Therefore, $\det(\mathbf{M}) = ab - c^2 = \det(\mathbf{\Lambda}) = \lambda_1 \times \lambda_2$, which, along with Equation 7.37, demonstrates the result in Equation 7.28.

Although the introduction of a cross-term $c \neq 0$ looks like only a minor complication, the generalization of Equation 7.27 to Equation 7.28 turns out to be very useful for practical applications. For example, in Chapter 5, we discussed methods to approximate integrals, including Laplace's method. For a function $g(x, y)$ of two variables, expressions like Equation 7.28 arise naturally from the two-dimensional Taylor series expansion about a maximum at (x_0, y_0). At the location of the maximum, the first derivatives are zero, so that

$$g(x, y) = g(x_0, y_0) + \frac{1}{2}\left((x-x_0)^2 \frac{\partial^2 g}{\partial x^2}\bigg|_{\substack{x=x_0 \\ y=y_0}} + (y-y_0)^2 \frac{\partial^2 g}{\partial y^2}\bigg|_{\substack{x=x_0 \\ y=y_0}} \right.$$
$$\left. + 2(x-x_0)(y-y_0) \frac{\partial^2 g}{\partial x \partial y}\bigg|_{\substack{x=x_0 \\ y=y_0}} \right) + \dots \tag{7.46}$$

The matrix elements a, b, and c in Equation 7.28 can be associated to the second derivatives in Equation 7.46 by

$$\mathbf{M} = \begin{bmatrix} a & c \\ c & b \end{bmatrix} = \begin{bmatrix} \dfrac{\partial^2 g}{\partial x^2} & \dfrac{\partial^2 g}{\partial x \partial y} \\ \dfrac{\partial^2 g}{\partial x \partial y} & \dfrac{\partial^2 g}{\partial y^2} \end{bmatrix}. \tag{7.47}$$

There is a nonzero term c whenever the mixed partial derivative evaluated at the maximum (x_0, y_0) is nonzero, which is often the case. Finally, as discussed in texts on multivariable calculus, the conditions that a and $\det(\mathbf{M})$ are positive ensure that g has a maximum at the point (x_0, y_0).

Although Example 7.3 assumes a symmetric, 2×2 matrix $\mathbf{M}$, all of the linear algebra steps remain valid for higher dimensions. For example, the three-dimensional integral

$$J_3 = \int_{z=-\infty}^{\infty} \int_{y=-\infty}^{\infty} \int_{x=-\infty}^{\infty} e^{-\frac{1}{2}Q_3(x,y,z)} dx \, dy \, dz \tag{7.48}$$

can be evaluated in a similar manner by diagonalizing the 3×3 symmetric matrix $\mathbf{M}_3$:

$$Q_3(x, y, z) = \begin{bmatrix} x & y & z \end{bmatrix} \begin{bmatrix} a & c & d \\ c & b & e \\ d & e & f \end{bmatrix} \begin{bmatrix} x \\ y \\ z \end{bmatrix} = \mathbf{x}^T \mathbf{M}_3 \mathbf{x}. \tag{7.49}$$

Steps analogous to those used in Example 7.3 yield

$$I_3 = \frac{(2\pi)^{3/2}}{\sqrt{\det \mathbf{M}_3}}, \tag{7.50}$$

provided that all of the eigenvalues of $\mathbf{M}_3$ are positive. For the three-dimensional case, there is also a geometric interpretation. The equation $Q_3(x, y, z) = 1$, which can be expanded using Equation 7.49, represents an ellipsoid. We find a rotated set of coordinates (u, v, w) that are aligned with its principal axes, so the cross-terms in Equation 7.49 drop out.

Although the geometric interpretation is no longer as clear, the same procedure can be used to evaluate the N-dimensional integral

$$I_N = \int\limits_{x_N = -\infty}^{\infty} \cdots \int\limits_{x_2 = -\infty}^{\infty} \int\limits_{x_1 = -\infty}^{\infty} e^{-\frac{1}{2}Q_N(x_1, x_2, \ldots, x_N)} \, dx_1 \, dx_2 \ldots dx_N = \frac{(2\pi)^{N/2}}{\sqrt{\det \mathbf{M}_N}}, \tag{7.51}$$

again, provided that all of the eigenvalues of $\mathbf{M}_N$ are positive. Equation 7.51 is a general and useful result because, for large values of N, the definite integral becomes too computationally complex to evaluate, even with numerical methods.

7.4   GENERAL YET SIMPLE: EULER'S IDENTITY

The previous examples illustrate that as an equation is generalized, it tends to become less concrete and more abstract. Often, generalized expressions are more complicated as well, but occasionally, they can be simpler than an embedded special case. A comparison of the famous identities of De Moivre and Euler provides such an example. The formula involving trigonometric functions and $i = \sqrt{-1}$,

$$(\cos\theta + i\sin\theta)^n = \cos n\theta + i\sin n\theta, \tag{7.52}$$

is named after the French-born mathematician Abraham De Moivre. Although Equation 7.52 is certainly a remarkable result, it can be considered to be a special case of an even more remarkable result,

$$e^{i\theta} = \cos\theta + i\sin\theta, \tag{7.53}$$

because $\left(e^{i\theta}\right)^n = e^{in\theta}$. Equation 7.53 is known as Euler's identity, after Leonhard Euler, the great Swiss mathematician.

Many very important equations are both simple and general. Familiar examples include $F = ma$ and $E = mc^2$. Exercise 3.7 showed how appropriate variable substitution simplified van der Waals' equation of state, while increasing its generality. The generality and simplicity of Euler's identity, Equation 7.53, help to explain its central importance in physics, engineering, and other fields. The next example illustrates how that identity can simplify the solution of a differential equation that arises in many mechanical engineering applications.

Consider a model for a mechanical system comprising a mass m attached to a spring with force constant k, i.e., a harmonic oscillator. As described in Section 5.5, any one-dimensional mechanical system that has a minimum in its potential energy function can be approximated by this configuration. The system is driven by a time-dependent, sinusoidal force $F(t) = F_0 \cos\omega t$. This form for the force is not as specialized as it looks because, as discussed in Example 4.4, any more general, periodic driving function can be decomposed into its Fourier series components. The displacement of a mass m from its equilibrium location is given by $x(t)$, and the damping forces that oppose motion are modeled to be proportional to the velocity. Then, according to Newton's second law of motion $F = ma$, the displacement $x(t)$ of the damped oscillator can be described by the second-order, linear differential equation

$$m\frac{d^2x}{dt^2} + b\frac{dx}{dt} + kx = F_0 \cos\omega t, \qquad (7.54)$$

where b is the damping constant. As discussed in most undergraduate texts in classical mechanics, Equation 7.54 has transient solutions that damp out after a period of time and also a steady-state solution that persists. Euler's identity facilitates finding the solution for the steady-state motion.

EXAMPLE 7.4

Although all of the quantities in Equation 7.54 are real, consider a generalized version of the equation in which the driving force is complex. We can write the complex force as

$$\hat{F}(t) = F_0 e^{i\omega t} = F_0(\cos\omega t + i\sin\omega t), \qquad (7.55)$$

so that the real, physical force is $F(t) = \text{Re}[\hat{F}(t)]$. We will also allow the displacement to be complex, denoted as $\hat{x}(t)$, so that the real, physical displacement is given by $x(t) = \text{Re}[\hat{x}(t)]$.

(a) Write a real-to-complex generalization of Equation 7.54. Solve for the steady-state displacement in this generalized equation. This can be accomplished by using the trial solution $\hat{x}(t) = Ae^{i\omega t}$, which permits cancellation of the time-dependent factors. Here, A is a complex constant. Does the mass always oscillate in phase with the driving force?

(b) Examine the "resonance" condition for $x(t)$, that is, the value of the driving angular frequency ω, where the amplitude of the displacement is a maximum.

ANSWER

The generalized version of Equation 7.54 is

$$m\frac{d^2\hat{x}}{dt^2} + b\frac{d\hat{x}}{dt} + k\hat{x} = F_0 e^{i\omega t}. \tag{7.56}$$

This type of differential equation is typically solved by substituting the trial solution $\hat{x}(t) = Ae^{i\omega t}$ into Equation 7.56. Cancellation of the common factor $e^{i\omega t}$ and solving for the complex value A yields

$$A = \frac{F_0}{\left(k - m\omega^2\right) + ib\omega}. \tag{7.57}$$

The complex amplitude A can be decomposed into its magnitude and phase according to $A = |A|e^{i\theta}$. From Equation 7.57, we find

$$|A| = \frac{F_0}{\sqrt{\left(k - m\omega^2\right)^2 + (b\omega)^2}} \tag{7.58}$$

and

$$\angle A = \theta = \arctan\left(\frac{b\omega}{m\omega^2 - k}\right). \tag{7.59}$$

The physical displacement is real and can be found by calculating $x(t) = \text{Re}[Ae^{i\omega t}]$, which yields

$$x(t) = |A|\cos(\omega t + \theta) = \frac{F_0}{\sqrt{\left(k - m\omega^2\right)^2 + (b\omega)^2}}\cos(\omega t + \theta), \tag{7.60}$$

with the phase shift θ given by Equation 7.59. The displacement of the mass oscillates with the same frequency ω as the driving force, but because $\theta \neq 0$, its phase can lag. Note that Equation 7.59 contains an arctangent function, and its denominator is zero when $\omega = \sqrt{k/m}$. The arctangent function is periodic and repeats every 180^0 or π radians, so there is some freedom to choose which of its branches to use. It is convenient to choose the range $-180^0 < \theta < 0$, with $\theta = -90^0$ when $\omega = \sqrt{k/m}$. With that choice, the phase shift θ varies continuously from 0 to -180^0 as the driving frequency ω varies from 0 to ∞. Notice that for very small driving frequencies, the response can "keep up," i.e., $\theta \approx 0$, which is what we would expect, intuitively.

(b) The magnitude of the amplitude depends on the frequency of the driving force. Using Equation 7.58 to calculate $\dfrac{d|A|}{d\omega}$ and setting the result equal to zero, we find that the amplitude reaches a <u>maximum</u> called resonance when the angular frequency $\omega_{\text{resonant}} = \sqrt{(k/m) - b^2/2m^2}$, provided that the damping is small, i.e., $b^2 < 2km$. At the resonant frequency, the amplitude of the motion is

$$|A|_{\text{resonant}} = \frac{F_0}{b\sqrt{\dfrac{k}{m} - \dfrac{b^2}{4m^2}}}, \tag{7.61}$$

while the phase shift is $\theta = -\arctan(2m\omega_{\text{resonant}}/b)$.

The use of the complex exponential in Example 7.4 greatly simplifies the algebraic manipulation required to solve the problem. The interested reader can verify the simplification by attempting to solve Equation 7.54 directly with a purely real, trial function such as $x(t) = B \cos(\omega t + \phi)$. Although the generalization to a complex displacement $\hat{x}(t)$ might seem abstract, this physical problem is particularly well suited to this approach, because the important results are characterized by an amplitude and a phase angle, as illustrated by Equations 7.58 and 7.59. A situation for which it is easier to solve a more general problem is an example of what is sometimes called the "inventor's paradox."

Generalizing an equation can save effort in other ways too. For example, the following differential equation describes a resistor–inductor–capacitor (RLC) electrical circuit that is driven by a sinusoidal voltage:

$$L\frac{d^2Q}{dt^2} + R\frac{dQ}{dt} + \frac{Q}{C} = V_0 \cos\omega t. \tag{7.62}$$

Notice that the structure of Equation 7.62 is identical to the structure of Equation 7.54. This allows us to set up an electromechanical analogy. We need only to make the proper identifications: displacement x with electrical charge Q, mass m with inductance L, etc. We then can obtain the solution of one problem directly from the solution of the other. The equivalent structures of Equations 7.54 and 7.62 can also aid prototyping efforts in the laboratory. For example, it might be much easier to build an RLC circuit than the equivalent damped mechanical oscillator.

7.5   WHEN TO TRY TO GENERALIZE

The previous example problems illustrate several techniques that are used to generalize equations. When would they be applied in practice?

Perhaps, the most common motivation to generalize is pragmatic: an equation is not general enough to model an engineering application or results of a scientific experiment, i.e., it is not sophisticated enough to describe the observed data. For example, a simplified model of an oscillator that has no damping can be obtained from Equation 7.54 by omitting the term proportional to b, or equivalently by setting $b = 0$. In this simplified case, Equation 7.61 indicates that the amplitude of motion becomes infinite at the resonant frequency, which cannot accurately model any physical system. Still, the simplified case does suggest that dramatic behavior will be observed near the resonant frequency, which is true. The failure of the simplified equation at the resonant frequency provides motivation to seek a more general model, i.e., one in which $b > 0$. In this way, the mathematical model is refined by using a more general equation.

An alternative to this pragmatic approach is to try to generalize equations whenever possible. This strategy is particularly useful for discovering new ideas in mathematics, and sometimes it can be applied to the physical sciences as well. Naturally, there is no guarantee that the generalized equation will be useful, but the history of mathematics and science illustrates that occasionally a valuable result can be found. This is illustrated by the work of Sir William Hamilton, an Irish mathematician and physicist.

In the early 1800s the equations of Newtonian mechanics were reformulated from the original version $F = ma$, which involves forces, to other versions based on potential and kinetic energy. We are not concerned with the details here, but one of the formulations is called the Hamilton–Jacobi equation, co-named after Carl Jacobi, a German mathematician. The Hamilton–Jacobi theory provides an interesting connection between the motion of objects (i.e., mechanics) and the results of geometric optics. The latter describes the paths of rays of light. The link is established because light obeys Fermat's principle of least time (mentioned in Chapter 5), and the mechanical system obeys a related principle of least action. Geometric optics, which is valid only for short wavelengths of light, is a special case of the more general theory of wave optics, which is valid at all wavelengths. In this context, "short" means that the wavelength of the light is small compared with the distance over which the index of refraction of the medium varies.

Hamilton conceivably could have generalized the Hamilton–Jacobi equation by making a link between mechanics and the more general wave optics. Perhaps then he could have continued to discover the Schrödinger equation.

Instead, that result was obtained almost a century later by Erwin Schrödinger, a German physicist. The Schrödinger equation is central to quantum mechanics, which remains among the most important scientific discoveries of the twentieth century.

This historical example not only illustrates the value of trying to generalize equations, but also highlights a fundamental difference between mathematics and the physical sciences. Although a generalization to the Schrödinger equation was mathematically available, there was no motivation for Hamilton to make such a leap. At that time, no experiment suggested the need to develop a "long-wavelength version" of Newtonian mechanics. Physical sciences are empirical, driven by experimental observation. In the absence of motivating data, the generalization was not made.

EXERCISES

(7.1)   Draw a diagram that provides a geometric interpretation of Equation 7.2. How would you generalize your geometric interpretation to describe $(x + y)^3$? Hint: if you have access to 27 cubes, such as children's building blocks, build a configuration to illustrate that $(1 + 2)^3 = 1^3 + (3 \times 1^2 \times 2) + (3 \times 1 \times 2^2) + 2^3 = 27$.

(7.2)   Use Equation 7.38 to show that the Jacobian determinant, i.e., the absolute value of

$$\det \begin{bmatrix} \dfrac{\partial u}{\partial x} & \dfrac{\partial u}{\partial y} \\ \dfrac{\partial v}{\partial x} & \dfrac{\partial v}{\partial y} \end{bmatrix}$$

is equal to 1. Consequently, we can change the variable in the two-dimensional integral without introducing any scale factors, that is, $dx\, dy = du\, dv$. Is this the expected result from a rotation of coordinate axes? What would you expect for the Jacobian determinant if the uv-coordinate system was stretched instead of rotated?

(7.3)

    (a) Show that Equations 7.15, 7.20, and 7.24 are dimensionally consistent. Assume each Cartesian coordinate is measured in meters.

    (b) Show that Equation 7.60 is also dimensionally consistent. Does the equation make sense for the limiting case $\omega = 0$?

(7.4)

    (a) Verify Equation 7.53 by considering the power series:

$$e^z = 1 + z + \frac{z^2}{2!} + \dots,$$

$$\cos z = 1 - \frac{z^2}{2!} + \frac{z^4}{4!} + \dots,$$

$$\sin z = z - \frac{z^3}{3!} + \frac{z^5}{5!} + \dots$$

(b) Verify that both sides of Equation 7.53 satisfy the differential equation $f''(z) + f(z) = 0$, with the initial conditions $f(0) = 1$ and $f'(0) = i$.

(c) Use Equation 7.53 to verify the identity $\sin^2\theta + \cos^2\theta = 1$. Hint: use $e^{i0} = 1$.

(7.5) Use Equation 7.53 to verify the trigonometric identity $\cos(a + b) = \cos a \cos b - \sin a \sin b$ by considering the real part of $e^{ia} \times e^{ib}$.

(7.6) Consider a further generalization of the binomial formula, Equation 7.7

$$(x_1 + x_2 + \dots + x_q)^n = \sum \frac{n!}{n_1! n_2! \dots n_q!} x_1^{n_1} x_1^{n_2} \dots x_1^{n_q},$$

where the sum is taken over all the nonnegative integers so that $n_1 + n_2 + \dots + n_q = n$. Show that this multinomial formula reduces the binomial formula when $q = 2$. This is an example of generalizing an equation to more variables or degrees of freedom.

(7.7) For integers $n \geq r$, give an argument based on combinatorics why $\binom{n}{r} = \binom{n}{n-r}$. Then, show that the result is true using the definition given by Equation 7.6.

(7.8) Equation 7.10 is related to the identity

$$\sum_{k=0}^{n} \binom{b}{k} \times \binom{g}{n-k} = \binom{b+g}{n}, \quad b \geq n \text{ and } g \geq n.$$

Give a combinatoric argument for this result by considering a group of b boys and g girls and counting how many different ways the group of n children can be chosen.

(7.9) Another common way to generalize is to go from a vector equation to a "vector operator" equation. Recall the definition of a vector dot and cross products

$$\vec{a}\cdot\vec{b} = a_x b_x + a_y b_y + a_z b_z \quad \text{and} \quad \vec{a}\times\vec{b} = \det\begin{bmatrix} i & j & k \\ a_x & a_y & a_z \\ b_x & b_y & b_z \end{bmatrix},$$

where $\hat{i}, \hat{j},$ and $\hat{k}$ are unit length vectors in the $x, y,$ and z directions, respectively.

(a) Show that for any three vectors

$$\vec{a}\times(\vec{b}\times\vec{c}) = \vec{b}(\vec{a}\cdot\vec{c}) - \vec{c}(\vec{a}\cdot\vec{b}). \tag{7.63}$$

(b) The operator "del" is defined as $\nabla = \hat{i}\dfrac{\partial}{\partial x} + \hat{j}\dfrac{\partial}{\partial y} + \hat{k}\dfrac{\partial}{\partial z}$. Del is sometimes called a vector operator because it has three components like a regular vector, and its partial derivatives operate on a function. Suppose $\vec{f}$ is a vector function, i.e., $\vec{f}(x, y, z) = \hat{i}\, f_x(x, y, z) + \hat{j}\, f_y(x, y, z) + \hat{k}\, f_y(x, y, z).$ Then, show that

$$\nabla\times(\nabla\times\vec{f}) = \nabla(\nabla\cdot\vec{f}) + (\nabla\cdot\nabla)\vec{f} = \nabla(\nabla\cdot\vec{f}) + \nabla^2\vec{f}. \tag{7.64}$$

(Recall $\nabla\times\vec{f}$ is called the curl of $\vec{f}$, $\nabla\cdot\vec{f}$ is called the divergence of $\vec{f}$, and $\nabla^2\vec{f}$ is called the Laplacian of $\vec{f}$.)

In what ways is Equation 7.64 a generalization of Equation 7.63? Note that the order of the factors in the last term does not matter in Equation 7.63, but does matter in Equation 7.64.

(7.10) If we accept that translations and rotations do not change the area of a triangle, then we can use properties of determinants to demonstrate that Equation 7.15 is a general result, i.e., the special cases considered in Example 7.2 show its validity "without loss generality."

(a) The value of a determinant is unchanged if one row (or column) is multiplied by a factor and added to another row (or column). Use this property to show that

$$\det\begin{pmatrix} x_1 & y_1 & 1 \\ x_2 & y_2 & 1 \\ x_3 & y_3 & 1 \end{pmatrix} = \det\begin{pmatrix} 0 & 0 & 1 \\ x_2-x_1 & y_2-y_1 & 0 \\ x_3-x_1 & y_3-y_1 & 0 \end{pmatrix} \equiv \det\begin{pmatrix} 0 & 0 & 1 \\ x_2' & y_2' & 0 \\ x_3' & y_3' & 0 \end{pmatrix}.$$

Geometrically, this first step translates vertex 1 to the origin of the coordinate system. Then, show that

$$\det\begin{pmatrix} 0 & 0 & 1 \\ x_2' & y_2' & 0 \\ x_3' & y_3' & 0 \end{pmatrix} = \det\begin{pmatrix} 0 & 0 & 1 \\ x_2' & y_2' & 1 \\ x_3' & y_3' & 1 \end{pmatrix}.$$

(b) Once vertex 1 coincides with the origin, we can apply Equation 7.43. Specifically,

$$\det\begin{pmatrix} 0 & 0 & 1 \\ x_2' & y_2' & 1 \\ x_3' & y_3' & 1 \end{pmatrix} = \det\left\{ \begin{pmatrix} 0 & 0 & 1 \\ x_2' & y_2' & 1 \\ x_3' & y_3' & 1 \end{pmatrix} \times \begin{pmatrix} \cos\theta & -\sin\theta & 0 \\ \sin\theta & \cos\theta & 0 \\ 0 & 0 & 1 \end{pmatrix} \right\},$$

where the matrix on the right is an orthogonal rotation matrix. Show that the determinant of the rotation matrix equals to 1. Then, show that choosing the rotation angle $\theta = \arctan(y_2'/x_2')$ brings vertex 2 to the x-axis, as in Fig. 7.1. That is, show that

$$\begin{pmatrix} 0 & 0 & 1 \\ x_2' & y_2' & 1 \\ x_3' & y_3' & 1 \end{pmatrix} \times \begin{pmatrix} \cos\theta & -\sin\theta & 0 \\ \sin\theta & \cos\theta & 0 \\ 0 & 0 & 1 \end{pmatrix} = \begin{pmatrix} 0 & 0 & 1 \\ x_2'' & 0 & 1 \\ x_3'' & y_3'' & 1 \end{pmatrix}.$$

Discussion: The general problem has been reduced to the simplified special case described in Example 7.2a. Because these properties of determinants are true regardless of the dimension of the matrix, we can generalize the methods of this exercise to even higher dimensions. For example, in three dimensions, we can translate the vertex 1 of an arbitrary tetrahedron to the origin, and then apply a three-dimensional rotation so that vertices 2 and 3 also lie in the xy-plane. This reduces the general problem to a special case that is covered by Equation 7.22. This method can be further generalized to higher dimensions to validate the general expression for hyper-volume in Equation 7.25, provided we accept several assumptions. First, the N-dimensional hyper-volume of a generalized tetrahedron with $N + 1$ vertices is equal to the product of its "height" and the (N-1)-dimensional hyper-area of its "base," divided by the dimension N. Second, the hyper-area of its base can be found by rotating the object characterized by N vertices into an (N-1)-dimensional space and then finding its (N-1)-dimensional hyper-volume. The height is the distance of the remaining vertex along the perpendicular dimension. Returning to the example of the three-dimensional tetrahedron, the base corresponds to a triangle in a two-dimensional space, and the height is the z-coordinate of the fourth vertex.

(7.11) Pascal's triangle is a construction that can generate the binomial coefficients given by Equation 7.6. It contains many interesting patterns that can motivate general expressions. The first row is given by

$$\binom{0}{0} = 1.$$

Entries in subsequent rows of Pascal's triangle are generated with the recursion relation

$$\binom{n}{k-1}+\binom{n}{k}=\binom{n+1}{k},$$

i.e., simply summing the two entries directly above it in the triangle.

Another interesting property is that the sum along any diagonal is equal to an entry in the next row. For example, consider the entries in boldface type:

$$
\begin{array}{ccccccccccc}
& & & & & 1 & & & & & \\
& & & & \mathbf{1} & & 1 & & & & \\
& & & 1 & & 2 & & 1 & & & \\
& & 1 & & 3 & & \mathbf{3} & & 1 & & \\
& 1 & & 4 & & 6 & & \mathbf{4} & & 1 & \\
1 & & 5 & & 10 & & \mathbf{10} & & 5 & & 1
\end{array}
$$

that is, $1 + 2 + 3 + 4 = 10$. Write a general expression involving the sum of binomial coefficients that expresses this property, and verify it for the special case shown earlier, $\binom{5}{3} = 10$, and for a special case of your own choosing.

(7.12)

(a) Verify that Equation 7.15 gives the correct answer for the area of a triangle if two of its vertices lie on the same point.

(b) If the labels of two vertices of a triangle are interchanged (e.g., vertex 1 is relabeled vertex 2, and vice versa), what happens to the area of the triangle? Verify that Equation 7.15 displays the proper symmetry under this operation.

(c) Verify that Equation 7.15 also displays the proper symmetry if the triangle is reflected about the x-axis, i.e., $(x_1, y_1) \rightarrow (x_1, -y_1)$, and similarly for the other two vertices. What happens if the triangle is reflected about the origin?

(7.13) The concept of the nth derivative of a function can be generalized to fractional values of n by using the gamma function, which is related to the factorial by $\Gamma(n) = (n-1)!$

(a) Consider the function $y = x^k$. Show that the nth derivative of y is

$$\frac{d^n y}{dx^n} = k(k-1)\ldots(k-n+1)x^{k-n} = \frac{k!}{(k-n)!}x^{k-n}. \tag{7.65}$$

Use the relationship $\Gamma(n) = (n - 1)!$, to show that Equation 7.65 is equivalent to

$$\frac{d^n}{dx^n}x^k = \frac{\Gamma(k+1)}{\Gamma(k-n+1)}x^{k-n}. \tag{7.66}$$

(b) Verify Equation 7.65 for the special case $k = 5$ and $n = 3$.

(c) Use the fact that $\Gamma(1.5) = \sqrt{\pi}/2$ and Equation 7.66 to show that the one-half derivative of x is

$$\frac{d^{1/2}}{dx^{1/2}}x = 2\left(\frac{x}{\pi}\right)^{1/2}.$$

(d) Show that

$$\frac{d^{1/2}}{dx^{1/2}}\sqrt{x} = \frac{1}{2}\sqrt{\pi}.$$

(e) Use the results in parts (c) and (d) to show that

$$\frac{d^{1/2}}{dx^{1/2}}\left(\frac{d^{1/2}}{dx^{1/2}}x\right) = 1,$$

which is equal to the first derivative of x, as we would expect. Equation 7.66 can be further generalized to include complex values of n because the gamma function is single valued and analytic over the complex plane, except at the negative integers, where it diverges.

FURTHER READING

Goldstein H. 1980. *Classical Mechanics*, 2nd ed. Reading, MA: Addison-Wesley.

This book contains more detailed discussion of the Hamilton–Jacobi theory and its relation to quantum mechanics. The principal axis theorem is also discussed here in more detail.

Taylor JR. 2005. *Classical Mechanics*. Sausalito, CA: University Science Books.

Many other undergraduate texts on classical mechanics such as this book provide a more detailed discussion about the transient and steady-state solutions of the driven harmonic oscillator.

8

SEVERAL INSTRUCTIVE EXAMPLES

In the final chapter, we present several instructive examples. In Section 8.1, we present two examples that illustrate how a clever choice of coordinate system can simplify the equations needed to solve a problem. Then, in Section 8.2, we discuss how insight can be gained when it is discovered that the solution to a problem has unanticipated properties. The solution might be simpler or more complicated than expected, or it might suggest unexpected symmetry properties. In Section 8.3, we examine an indirect method of solving problems by starting with known solutions that apply to a broad class of problems. We illustrate this method with an example from the field of electrostatics. Specifically, we discuss an algorithm that generates expressions guaranteed to satisfy the two-dimensional version of Laplace's equation, even though it is not known beforehand to which problem the generated solution will correspond. Finally, in Section 8.4, we present examples from the fields of chaos theory and cosmology that illustrate the far-reaching ways in which equations can alter our understanding of the world.

8.1 CHOICE OF COORDINATE SYSTEM

Choosing the optimal coordinate system can greatly simplify finding the solution to a problem. Consider the simple problem of calculating the distance

Thinking About Equations: A Practical Guide for Developing Mathematical Intuition in the Physical Sciences and Engineering, by Matt A. Bernstein and William A. Friedman
Copyright © 2009 John Wiley & Sons, Inc.

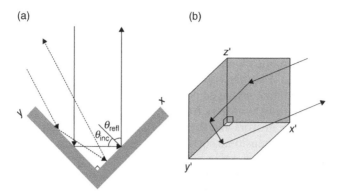

FIG. 8.1. A corner reflector directs light straight back in the direction from which it originates. (a) A two-dimensional corner reflector. (b) A three-dimensional corner reflector, also known as a corner cube.

between the points in space represented by the locations $\vec{p}_1 = (1, 1, 1)$ and $\vec{p}_2 = (1, 2, 3)$. Because the distance between two points is invariant under translation of the origin of the coordinate system, it is convenient to introduce a primed coordinate system, which has its origin at $\vec{p}_1$. In that coordinate system, $\vec{p}_1' = (0, 0, 0)$ and $\vec{p}_2' = (1-1, 2-1, 3-1) = (0, 1, 2)$. The distance between $\vec{p}_1$ and $\vec{p}_2$ is easily seen to be $\sqrt{1^2 + 2^2} = \sqrt{5}$ units. Translating the origin of the coordinate system is hardly necessary to solve this simple problem, but the following examples more fully demonstrate the advantages of a judicious choice of coordinate system.

A device called a *corner reflector* is designed to reflect a ray of light back in the exact opposite direction from which it came. This property is independent of the incident direction of the ray. Corner reflectors are used in optics as well as in the construction of safety devices such as the reflectors that make bicycles more visible at night. Corner reflectors that reflect back radar signals serve as navigation aids on ships and buoys. Of course, both radar waves and visible light are examples of electromagnetic waves, but the wavelength of visible light is approximately six orders of magnitude shorter.

Figure 8.1a illustrates a schematic diagram of a two-dimensional corner reflector. The incoming ray sequentially reflects off two perpendicular, planar mirrors. Regardless of the direction of the incoming ray, the direction of the outgoing ray is exactly reversed. The outgoing ray is offset, but its direction is antiparallel to the incident ray. This result can be verified (see exercise 8.1) by using the law of reflection, which is a principle from geometric optics that governs specular reflection, i.e., reflection from a smooth surface like a mirror. The law states that (1) the reflected ray remains in the plane defined by the incident ray and the normal to the surface at the point of incidence, and (2) the angle of incidence equals the angle of reflection,

$$\theta_{inc} = \theta_{refl}, \tag{8.1}$$

as depicted on Fig. 8.1a.

Figure 8.1b schematically shows a three-dimensional corner reflector, which is sometimes called a *corner cube*. A corner cube is constructed from three planar mirrors, each one arranged perpendicular to the other two. An incident ray of light (originating anywhere from one octant of solid angle) reflects once off each of the three mirrors and then emerges from the corner cube. Just as in the case of the two-dimensional corner reflector, the outgoing ray is antiparallel to the direction of the incoming ray. The verification of that result becomes complicated when repeatedly applying on Equation 8.1 for each reflection. A strategic choice of coordinate system, however, simplifies the problem greatly.

EXAMPLE 8.1

Show that the ray of light emerging from a corner cube travels in a direction that is antiparallel to its incoming direction, regardless of the specific direction of the incoming ray.

ANSWER

Suppose the direction of an arbitrary incoming ray of light is described by the three-dimensional unit vector $\hat{n}_0 = (n_{0x}, n_{0y}, n_{0z})$. One reasonable choice of coordinate system is to let the incident direction of the ray correspond to the x-direction, i.e., $\hat{n}_0 = (1, 0, 0)$. To show the desired result, we must consider the planes of the three mirrors in this coordinate system and trace the ray through its three reflections. We would then determine whether or not the direction of the emerging ray is reversed, i.e., $\hat{n}_3 = (-1, 0, 0)$, where the subscript indicates the number of times that the ray has been reflected.

Next, consider a second coordinate system that has its axes fixed to the three lines of intersection of the mirrors in the corner cube. We will call this choice the $x'y'z'$-coordinate system, and it is illustrated in Fig. 8.1b. In the second coordinate system, the direction of the incident ray is described by the unit vector $\hat{n}'_0 = (n'_{0x}, n'_{0y}, n'_{0z})$. Note that both $\hat{n}_0$ and $\hat{n}'_0$ represent the *same* direction in space but are expressed in two different coordinate systems. In general, the numerical values of the components of the unit vector $\hat{n}'_0$ differ from that in the first coordinate system, $\hat{n}_0 = (1, 0, 0)$. The components of the unit vectors in the two coordinate systems are related by a 3×3 rotation matrix $\mathbf{R}$:

$$\hat{n}'_0 = \mathbf{R}\hat{n}_0. \tag{8.2}$$

It is relatively easy to describe a reflection from any of the three mirrors in the corner cube in the primed coordinate system, which is a substantial advantage for this problem. Consider the first reflection from the mirror, which, as depicted in Fig. 8.1b, lies in the $x'z'$-plane based on our definition of the primed coordinate system. The only effect of this reflection is to reverse the direction of the ray along the y'-axis. Therefore, the unit vector describing its propagation direction after one reflection is

$$\hat{n}_1' = (n_{0x}', -n_{0y}', n_{0z}').\tag{8.3}$$

The next reflection is from the mirror in the $y'z'$-plane, so after two reflections,

$$\hat{n}_2' = (-n_{0x}', -n_{0y}', n_{0z}').\tag{8.4}$$

After the final reflection, the unit vector describing the direction of the ray emerging from the corner cube is

$$\hat{n}_3' = (-n_{0x}', -n_{0y}', -n_{0z}').\tag{8.5}$$

From Equation 8.5, we conclude that

$$\hat{n}_3' = -\hat{n}_0'.\tag{8.6}$$

We now return to the unprimed xyz-coordinate system and consider how $\hat{n}_3$ relates to $\hat{n}_0$. From the properties of vectors, it is expected that if the emerging ray is antiparallel to the incident ray as described in one coordinate system, it must be antiparallel in any other coordinate system as well. This result can be verified formally with the use of Equations 8.2 and 8.6,

$$\hat{n}_3 = \mathbf{R}^{-1}\hat{n}_3' = \mathbf{R}^{-1}(-\hat{n}_0') = \mathbf{R}^{-1}(-\mathbf{R}\hat{n}_0) = -(\mathbf{R}^{-1}\mathbf{R})\hat{n}_0 = -\hat{n}_0,\tag{8.7}$$

where $\mathbf{R}^{-1}$ is the inverse of the rotation matrix. Equation 8.7 confirms that the corner cube has the desired property.

Describing the three reflections in a coordinate system that is tied to the corner cube, rather than to the direction of the incident ray, provided a great deal of simplification for this problem.

The next example examines the mechanics of a rope supported by pegs and again demonstrates the advantages of a wise choice of the coordinate system. This example also illustrates the parametric substitution for solving differential equations, which was introduced in Chapter 2. It also illustrates a change in the number of real solutions that occurs at a transition point, which was discussed in Chapter 3.

EXAMPLE 8.2

Consider a flexible but non-stretchable rope that has uniform mass per unit length μ. It therefore has weight per unit length $w = \mu g$, where g is the acceleration due to gravity. If we drape the rope over a single, smooth peg, we can anticipate what will happen. We expect the rope to balance on the peg only when the lengths of the rope on each side of the peg are equal. This balance (i.e., mechanical equilibrium) can be achieved for any length of rope. If, however, one side is longer than the other, we expect the force of gravity to drag the entire rope to that side.

What will happen to the rope if there are two pegs separated in the horizontal and vertical directions, as shown in Fig. 8.2?

(a) Find the shape that the rope takes, assuming that it is held fixed at both pegs.

(b) Find the shape that the rope assumes if it is free to slide over both pegs.

(c) Consider the case where the rope is free to slide over both pegs and the heights of the two pegs are equal. Is there a minimum length L of rope that is necessary to achieve mechanical equilibrium? What configuration will the equilibrium cases have?

ANSWER

As the rope sags between the two pegs, it will define a plane that we label with the x- and y-axes. As is customary, we will take the x-direction to be horizontal

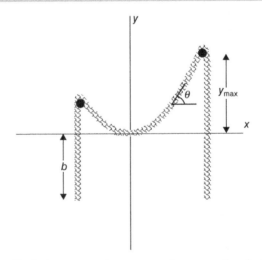

FIG. 8.2. A rope freely hangs over two pegs and assumes the shape of a catenary. The origin of the coordinate system $(0, 0)$ coincides with the lowest portion of the rope between the two pegs.

and the y-direction to be vertical. A convenient choice for the *origin* of this coordinate system is the lowest point of the rope where it sags between the two pegs. This choice for the origin greatly simplifies the problem. Outside of the pegs, the rope will hang vertically downward. This is shown in Fig. 8.2, where length b measures the portion of rope that extends below the x-axis and y_{max} measures the height of the right-hand peg above the x-axis.

(a) Consider separately the two ascending portions of rope on either side of the origin. Assume that both of these portions of the rope are stationary and hence in mechanical equilibrium under the influence of the gravitational force on the rope. Let the horizontal force on the rope at the right-hand peg be T_H. The horizontal component of the tension (the force on the end of any segment of rope) at the lowest point (origin) must also be T_H to maintain equilibrium. In fact, because the gravitational force is purely vertical, the horizontal component of the tension on the rope must have the constant value T_H everywhere in between the two pegs, in order to maintain equilibrium.

Suppose that between the pegs, the rope assumes a shape described by the function $y(x)$. At each point, the orientation of the rope makes an angle θ measured from the horizontal. The angle θ varies along the path of the rope and is related to the derivative at each point by

$$\frac{dy}{dx} = \tan\theta. \tag{8.8}$$

Consider a segment of rope of length s that starts at the origin. The weight of this segment is equal to ws. For the segment to be in equilibrium, its weight must be balanced by the y-component of the tension force T_y. A rope can only sustain a tension force that is tangential to its shape, or in other words, the direction of the tension on any short segment of rope is collinear to the rope segment itself. Decomposing the tension into its x- and y-components yields

$$T_x = T\cos\theta = T_H \tag{8.9}$$

and

$$T_y = T\sin\theta = ws. \tag{8.10}$$

Equation 8.10 states that the vertical force component of the tension balances the force of gravity. Eliminating T from Equations 8.9 and 8.10 and then combining the results with Equation 8.8 yields

$$\tan\theta = \frac{dy}{dx} = \frac{ws}{T_H}. \tag{8.11}$$

An infinitesimal length ds along the rope can be expressed in terms of its x- and y-differentials by

$$ds = \sqrt{dx^2 + dy^2} = dx\sqrt{1 + \left(\frac{dy}{dx}\right)^2} \tag{8.12}$$

or from Equation 8.12 in terms of derivatives,

$$\frac{ds}{dx} = \sqrt{1 + \left(\frac{dy}{dx}\right)^2}. \tag{8.13}$$

Equations 8.11 and 8.13 yield the second-order differential equation

$$\frac{d^2y}{dx^2} = \frac{d}{dx}\left(\frac{dy}{dx}\right) = \left(\frac{w}{T_H}\right)\frac{ds}{dx} = \left(\frac{w}{T_H}\right)\sqrt{1 + \left(\frac{dy}{dx}\right)^2}. \tag{8.14}$$

Equation 8.14 governs the sag of the rope. It is of the form of Equation 2.62, which was solved in Example 2.3 using parametric substitution. The boundary conditions used in Example 2.3 also apply directly to this problem, so the solution can be obtained directly from Equation 2.69:

$$y(x) = \frac{T_H}{w}\left[\cosh\left(\frac{wx}{T_H}\right) - 1\right]. \tag{8.15}$$

Equation 8.15 represents the shape assumed by any naturally hanging flexible rope and is known as a *catenary*. This term is derived from the Latin word for chain, as are some other English words like "concatenate."

(b) We now let the rope slide freely over the pegs and extend the analysis to the entire rope, including the vertical rope segments outside of the pegs. The tension T calculated in part (a) now refers to any point along the rope that is in between the pegs. Equation 8.9 is equivalent to

$$T = T_H \sec\theta. \tag{8.16}$$

The secant of the angle θ can be related to the position x along the rope by a trigonometric identity and Equation 8.8:

$$\sec\theta = \sqrt{1 + \tan^2\theta} = \sqrt{1 + \left(\frac{dy}{dx}\right)^2}. \tag{8.17}$$

Equation 8.17 and the derivative $y'(x)$ calculated from Equation 8.15 yield

$$\sec\theta = \sqrt{1 + \sinh^2\left(\frac{wx}{T_H}\right)} = \cosh\left(\frac{wx}{T_H}\right). \tag{8.18}$$

Using Equations 8.15, 8.16, and 8.18, the tension T along the rope can be expressed in terms of the rise y and the horizontal component of the tension,

$$T = wy + T_H. \tag{8.19}$$

Equation 8.19 gives the correct answer at $y = 0$, where we previously established that $T = T_H$. At the height of the right-hand peg, whose vertical position has been labeled y_{max} in Fig. 8.2, the tension increases to $T = T_H + wy_{max}$. The additional contribution to the tension gained in going from the origin to the point y_{max} is wy_{max}. This increase is simply equal to the weight of a rope of length y_{max}. An analogous situation occurs at the left peg as well, although the vertical height of that peg is arbitrary and need not be y_{max}. At equilibrium, the total tension T in the rope at the location of each peg is balanced by the weight of the portion of the rope that hangs down vertically outside of that peg. Referring to Fig. 8.2, at the location of the right-hand peg we have

$$T = wy_{max} + wb. \tag{8.20}$$

Combining Equations 8.19 and 8.20 yields

$$T_H = wb. \tag{8.21}$$

The same analysis for the horizontal component of the tension T_H at the location of the left peg shows that Equation 8.21 holds on that side also. Consequently, at equilibrium, the additional length of rope b must be equal on the right- and left-hand sides. Just as in the case of a rope draped over a single, smooth peg, the left and right ends of the rope must be at the same height for equilibrium.

We next explore the influence of the horizontal placement of the pegs. Again, we concentrate on the right peg and label its horizontal location as x_{max}. Then, combining Equations 8.15 and 8.21 yields

$$y_{max} = b\left[\cosh\left(\frac{x_{max}}{b}\right) - 1\right]. \tag{8.22}$$

Equation 8.22 relates the equilibrium extension b to the location of the peg (x_{max}, y_{max}). We can solve numerically for the value of b from Equation 8.22, but to gain insight, it is helpful to examine some limiting cases as well. For $b \gg x_{max}$, applying the expansion $\cosh u = 1 + u^2/2 + \ldots$ yields

$$y_{max} \approx \frac{x_{max}^2}{2b}. \tag{8.23}$$

Equation 8.23 states that as b grows very large, y_{max} must be small. In this limit, the catenary becomes very shallow, which is expected due to the strong pull of weight of the long vertical sections of rope on either side of the peg. In the other limit, for $b \ll x_{max}$, we obtain

$$y_{max} \approx \frac{b}{2} e^{x_{max}/b}, \tag{8.24}$$

which suggests that for small values of b, y_{max} increases exponentially. This case also requires a very long rope.

(c) Next, we consider the equilibrium restrictions on the total length of the rope L. The results are reminiscent of the problems presented in Chapter 3 that dealt with transition points in the number of real roots to equations. For simplicity, we will assume that the two pegs are at the same height and then write an expression for the total length of rope L required for the equilibrium. Using the expression for ds in Equation 8.12, we find

$$L = 2 \left(y_{max} + b + \int_0^{x_{max}} \sqrt{1 + \left(\frac{dy}{dx}\right)^2} \, dx \right). \tag{8.25}$$

Equation 8.25 can be evaluated by using Equations 8.18, 8.21, and 8.22, which yields

$$L = 2 \left[b \cosh\left(\frac{x_{max}}{b}\right) + b \sinh\left(\frac{x_{max}}{b}\right) \right] = 2b e^{x_{max}/b}. \tag{8.26}$$

Figure 8.3 shows a plot of Equation 8.26. Because L represents a length, it must have a positive value to be physically meaningful. Therefore, from Equation 8.26, it is necessary that b is also positive. The right-hand side of Equation 8.26 has a minimum at $b = x_{max}$, which is readily found by taking the derivative of $2b \exp(x_{max}/b)$ with respect to b and setting the result to zero. The minimum value for the length of the rope occurs when

$$L_{crit} = 2b \exp(1) \approx 5.437b. \tag{8.27}$$

There is no solution for b that can maintain equilibrium when $L < L_{crit}$. As can be appreciated from Fig. 8.3, for $L > L_{crit}$, there are two solutions: one for $b > x_{max}$ and one for $b < x_{max}$. For $b \gg x_{max}$, expanding the exponential in Equation 8.26 yields $L \approx 2(b + x_{max})$. This result is the expected solution when the rope is taut, so the dip between the two pegs is very shallow. For this same length of rope, however, there will be a second solution where the dip is great and the approximation in Equation 8.24 holds.

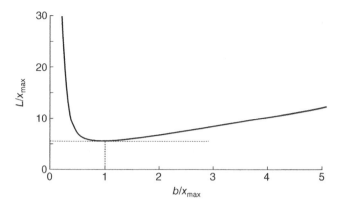

FIG. 8.3. A plot of Equation 8.26, indicating a minimum length of rope is required for the configuration to be stable. Note that the plot is dimensionally homogeneous because the quantities represented on both axes are dimensionless.

In this idealized situation, there are three cases for mechanical equilibrium, depending on the length L of the rope. These cases describe zero, one, or two specific stable configurations for the rope. This result is another example of an equation with a transition point, as discussed in Chapter 3. A real rope, however, will experience some friction on the pegs so that configurations close to these ideal ones could be stable too.

Besides being the natural form taken by a rope or chain draped over two supporting points, the catenary shape is utilized frequently in architectural designs. Presumably, this is because of the strength and stability of the inverted (i.e., concave down) catenary. The Spanish Catalan architect Antoni Gaudi used catenaries in many striking structures that enliven the city of Barcelona to the present day. Constructed more recently, the Gateway Arch near the Mississippi River in the city of St. Louis, Missouri, is also built in the shape of a catenary. As discussed in Example 8.2, this shape virtually eliminates the sheer components of force within the structure. The tremendous internal forces due to the extreme weight of the Gateway Arch remain tangential along the entire length of the structure, thereby increasing its stability.

8.2 SOLUTION HAS UNEXPECTED PROPERTIES

Many mathematical problems have no analytic solution. Other problems do have analytic solutions, but they are algebraically very complicated. Occasionally, however, a great amount of cancellation or simplification occurs as we work toward a solution. In these cases, we might find the ultimate solution to be much simpler than what we had expected. This situation is clearly good fortune, but it can also suggest an opportunity to gain insight leading to the discovery of an alternative and simpler approach, as illustrated by the next example.

EXAMPLE 8.3

Two new houses are to be connected to the city water supply. The water main can be tapped anywhere, but should only be tapped once to supply both houses with independent lines in a v-shaped pattern, as shown in Fig. 8.4. The water main runs north–south, and the first and second houses are 100 and 180 m east, respectively, of the water main. The second house is 110 m north of the first. How far north of the first house should the water main be tapped so that the total length of pipeline is minimized? How many meters of pipeline is required?

ANSWER

We choose the coordinate system so that the water main lies along the y-axis and the first house lies on the x-axis. Then, coordinates (in meters) of the two houses are $(x_1, y_1) = (100, 0)$ and $(x_2, y_2) = (180, 110)$. If the water main is tapped at $(0, h)$, then from Fig. 8.4, the total length of the pipeline p is given by the sum of two hypotenuses:

$$p = \sqrt{x_1^2 + h^2} + \sqrt{x_2^2 + (y_2 - h)^2}. \tag{8.28}$$

All the symbols in Equation 8.28 represent lengths, and so they are positive quantities. The minimum value of p can be found by differentiating Equation 8.28 with respect to h, setting the result equal to zero, solving for h, and finally, verifying that the second derivative of p is positive at that value of h. Setting the derivative equal to zero yields

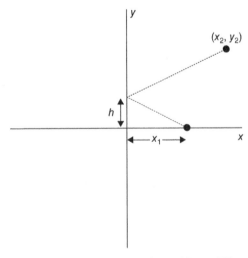

FIG. 8.4. Diagram for the water main problem of Example 8.3.

$$\frac{dp}{dh} = \frac{h}{\sqrt{x_1^2 + h^2}} - \frac{y_2 - h}{\sqrt{x_2^2 + (y_2 - h)^2}} = 0. \tag{8.29}$$

Eliminating the square roots from Equation 8.29 yields a fourth-order relationship for h:

$$h^2\left(x_2^2 + y_2^2 + h^2 - 2y_2 h\right) = \left(x_1^2 + h^2\right)\left(y_2^2 + h^2 - 2y_2 h\right). \tag{8.30}$$

Assembling terms in Equation 8.30 yields

$$h^4 - 2y_2 h^3 + h^2\left(x_2^2 + y_2^2\right) - \left[h^4 - 2y_2 h^3 + h^2\left(x_1^2 + y_2^2\right) - 2hy_2 x_1^2 + x_1^2 y_2^2\right] = 0. \tag{8.31}$$

The resulting coefficients of h^4 and h^3 are both zero in Equation 8.31, so the quartic equation reduces to a quadratic equation,

$$\left(x_2^2 - x_1^2\right)h^2 + 2x_1^2 y_2 h - y_2^2 x_1^2 = 0, \tag{8.32}$$

which has the solutions

$$h = \frac{-2x_1^2 y_2 \pm \sqrt{4x_1^4 y_2^2 + 4y_2^2 x_1^2\left(x_2^2 - x_1^2\right)}}{2\left(x_2^2 - x_1^2\right)}. \tag{8.33}$$

Equation 8.33 is a simpler solution than we might have expected from the quartic Equation 8.31, but it can be simplified even further. Cancelling terms in Equation 8.33, the square root term finally reduces to $2y_2 x_1 x_2$. Recalling that $x_2^2 - x_1^2 = (x_2 + x_1)(x_2 - x_1)$, the physically relevant solution for h in Equation 8.33 simplifies to

$$h = \frac{x_1}{x_1 + x_2} y_2. \tag{8.34}$$

Note the expression for h in Equation 8.34 is dimensionally correct, and the special cases $h|_{x_1=0} = 0$, $h|_{x_2=0} = y_2$, $h|_{y_2=0} = 0$, and $h|_{x_1=x_2} = y_2/2$ are all consistent with Fig. 8.4.

Substituting h from Equation 8.34 into Equation 8.28 yields the minimum length of pipeline that is required. Again, there is remarkable simplification, as we find that the complicated expression for the minimum length of pipe,

$$p_{\min} = \sqrt{x_1^2 + \left(\frac{x_1 y_2}{x_1 + x_2}\right)^2} + \sqrt{x_2^2 + \left(y_2 - \frac{x_1 y_2}{x_1 + x_2}\right)^2}, \tag{8.35}$$

is equal to

$$p_{min} = x_1 \sqrt{1 + \left(\frac{y_2}{x_1 + x_2} \right)^2} + x_2 \sqrt{1 + \left(\frac{x_2 y_2}{x_1 + x_2} \right)^2}, \tag{8.36}$$

which further reduces to

$$p_{min} = \sqrt{(x_1 + x_2)^2 + y_2^2}. \tag{8.37}$$

At this point, the problem is solved. With the numerical values specified, the results are $h \approx 39.3\,\text{m}$ and $p_{min} \approx 300.8\,\text{m}$.

At first, it might appear like purely good fortune that the quartic expression for h in Equation 8.31 ultimately has such a simple solution. The relative simplicity of the solution, however, is an indication that there might be a simpler way to look at the problem. This is often the case when a solution is simpler than expected.

Figure 8.5 shows what the simplicity of Equations 8.34 and 8.37 might be trying to tell us. If we reflect the location of the house 1 about the y-axis, we obtain a new point at $(-x_1, y_1) = (-100, 0)$. The total length of pipeline required is unchanged by the reflection. The shortest distance between house 2 and the reflection of house 1 about the y-axis is a straight line! Using precalculus level math, we can write down the result of Equation 8.37 by recognizing that it is the hypotenuse of the single, right triangle. Equation 8.34 also follows from setting up ratios of the lengths of sides of similar triangles:

$$\frac{h}{x_1} = \frac{y_2}{x_1 + x_2} \tag{8.38}$$

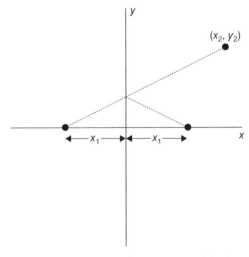

FIG. 8.5. The solution to Example 8.3 is simpler than might be expected because of the underlying reflection symmetry about the y-axis.

Perhaps, the clearest hint that there is a simpler approach to this problem comes from the reduction in the number of square root operations from two to one in the steps progressing from Equation 8.35 to Equation 8.37.

There is certainly nothing wrong with solving this problem using calculus. Instead, the purpose of Example 8.3 is to illustrate that the mathematical complexity of some problems can be greatly reduced after an insightful geometric or physical interpretation is found. In this example, the underlying symmetry of the isosceles triangle formed by the vertices $(-x_1, 0)$, $(0, h)$, and $(x_1, 0)$ simplifies the problem. Most real-world problems are very complicated, so unfortunately, it is rare for the solution to be simpler than expected. When it does occur, however, it is worthwhile exploring why, as valuable insight can be gained.

Example 8.3 is reminiscent of the method of *image charges*, which simplifies the solving of several problems in electrostatics involving a point electric charge q placed near a conducting surface. The method of image charges can often provide a simple, closed-form solution for the scalar electrostatic potential φ (measured in volts), without directly solving a differential equation subject to boundary conditions determined by the conducting surface and charge.

A basic application of the method of image charges is to solve for the electrostatic potential $\varphi(x, y, z)$ when a charge q is placed a distance a from a grounded conducting plane. "Grounded" means the potential φ is $0\,\mathrm{V}$ anywhere on the conducting plane. We will assume that the length and width of the conductor is much greater than the distance a, so that we can make the approximation that plane extends to infinity along its two dimensions.

We have all looked into a mirror and seen the image of a real object, such as our own face. We observe the image because the light rays reflected from the mirror's surface are identical to those that would have been produced by each point on our face if it were equivalently illuminated and located an equal distance on the opposite side of the mirror. This remarkable effect occurs because the mirror imposes exactly the right conditions on the reflected waves coming from each point of the real object to produce this illusion.

In the electrostatic problem, as the charge q is brought near the grounded conducting plane, charges of opposite sign are induced on the conductor so that the boundary condition $\varphi = 0$ is maintained. That is, if $q > 0$, the positive charge attracts the precise amount and distribution of negative charge from ground to the surface of the conductor so that $\varphi = 0$ there. Although the exact charge distribution on the conductor is quite complicated, we can emulate its entire effect in the region $x \geq 0$ by replacing it with a single point image charge (see Fig. 8.6). The simple configuration of q along with its image charge provides a function φ that satisfies the differential equation of electrostatics (i.e., Laplace's equation) in that region. The arrangement also satisfies the boundary condition at the surface of the conducting plane, so we know from the uniqueness theorem for boundary value problems that we have found the desired solution for the potential φ at any location with $x \geq 0$. In the following example, we work through some details and find the form of the scalar potential φ.

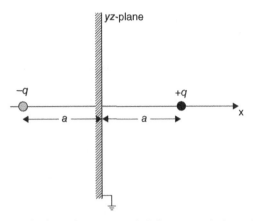

FIG. 8.6. A charge $+q$ is brought near an infinite, grounded conducting plane. The method of image charges provides a simple solution for the scalar potential.

EXAMPLE 8.4

Determine the scalar potential φ at all points to the right of an infinite planar conductor held at zero potential (i.e., grounded) when a point charge of strength q is placed a distance a from the conductor.

ANSWER

Choose the coordinate system so the conducting plane is located at $x = 0$, i.e., the conductor fills the yz-plane, and the charge is located at $x = a$ on the x-axis. Consider an image charge $-q$ located at $x = -a$ (Fig. 8.6). From symmetry, the potential is zero on the plane midway between the charges q and $-q$. This result follows from the cancellation of the potentials arising from each of the two point charges at every point on the plane. Therefore, we can replace the complicated charge distribution on the conducting plane with the single image charge. In the International System of Units (SI), the scalar potential at a distance r from a point charge q is $\varphi(r) = q/(4\pi\varepsilon_0 r)$, where ε_0 is the permittivity constant. The electrostatic potential at a point (x, y, z) is given by

$$\varphi(x, y, z) = \frac{1}{4\pi\varepsilon_0}\left(\frac{q}{\sqrt{(x-a)^2 + y^2 + z^2}} - \frac{q}{\sqrt{(x+a)^2 + y^2 + z^2}}\right), \quad (8.39)$$

provided that $x \geq 0$.

The electric field vector for points to the right of the conducting plane can be found from Equation 8.39 by calculating the gradient $\vec{E} = -\nabla\varphi$. After we calculate the electric field, we can apply Gauss's law to determine the actual spatial distribution of charge on the conducting plane. Not surprisingly, the final result is that the total charge induced on the conducting plane is equal to $-q$, but it is spread out in a complicated way on the conducting plane rather than being concentrated at a single point.

The method of image charges can be used to solve more complicated problems such as finding the electrostatic potential from multiple point charges placed near a conducting plane. This is accomplished by placing one image charge corresponding to each real point charge. To the extent that any charge distribution can be approximated by a set of point charges, the method can be further extended to describe any charged object brought near the conducting plane. This procedure is similar to the analysis of how the optical mirror produces a faithful image for objects of any size or shape. Another extension of the method of image charges is outlined in exercise 8.3, which involves a point charge near a grounded, conducting sphere.

Examples 8.3 and 8.4 describe instances where the solution is simpler than what we might have expected. On the other side of this same coin, sometimes a solution is more complicated or has other unanticipated properties. Consider the definite integral

$$\int_0^\infty \frac{\cos mx}{a^2 + x^2}\,dx = \frac{\pi e^{-ma}}{2a}. \tag{8.40}$$

Equation 8.40 provides a simple analytic result, which can be obtained from contour integration or by using properties of Fourier transforms. Based on Equation 8.40, we might expect that the very similar integral

$$I = \int_0^\infty \frac{\sin mx}{a^2 + x^2}\,dx \tag{8.41}$$

also has a simple, analytic form, but this is not the case. In this example, one difference is the integrand in Equation 8.40 is an even function, so the result is equal to one-half the integration from $-\infty$ to $+\infty$. We cannot make this same simplification in Equation 8.41 because its integrand is an odd function.

In Chapter 4, we discussed how symmetry can simplify equations and finding their solutions. If the equations display a symmetry, it might be reasonable to expect that their solutions will have the same symmetry. There are, however, important phenomena in the field of physics where this is not the case. In Chapter 3, we discussed ferromagnetism, which occurs in materials like iron below a critical temperature (i.e., the Curie temperature). The equations describing the interactions within the iron that result in the alignment of the atomic magnetic moments have spherical rotational symmetry.

The alignment can be induced by an external magnetic field that provides a direction of reference. That alignment, however, persists even after the external field is switched off and there is no longer a preferred direction. This situation indicates that the aligned state is indeed a solution to the equations that possess spherical symmetry, i.e., equations for which there is no preferred direction in space. Thus, in ferromagnetism, the underlying symmetry of the problem does *not* help us predict the properties of a solution.

The situation where a solution lacks the same symmetry possessed by the equations it solves is called *spontaneous symmetry breaking*. Associated with the broken symmetry is an order parameter, which was discussed in Chapter 3. In the case of ferromagnetism, the order parameter is the average magnetic moment. The value of this order parameter goes to zero when the spherical symmetry is restored, as the magnet is heated to a temperature hotter than the Curie temperature. Then, the "solution" is also independent of direction, and the symmetry of the underlying equations prevails.

For an even simpler example, consider a pencil balancing on its point. We all know that this condition is unstable and the pencil will fall down. The curious thing is that the physics that governs the stability of the pencil is symmetric under rotations about the vertical axis of the pencil. No particular direction in the horizontal plane is preferred or singled out. Yet, when the pencil does fall down, it goes in some particular direction. That solution, i.e., the eventual orientation of the fallen pencil, has lost the original rotational symmetry. That symmetry was spontaneously broken at the time of the fall.

The concept of spontaneous symmetry breaking is invoked in particle physics to account for the acquisition of mass by elementary particles. A confirmation of this mechanism is being sought in experiments performed at the Large Hadron Collider (LHC) particle accelerator, which is located near Geneva, Switzerland. In the 1960s, the English physicist Peter Higgs proposed the existence of a quantum field, now called the Higgs field. Very roughly, the Higgs field has properties analogous to the magnetization in a ferromagnetic material. The average value of the Higgs field is zero based on a particular symmetry, which is of a different nature than that of the familiar rotational symmetry that describes ferromagnetism. The breaking of this new symmetry is analogous to the average value of the magnetization differing from zero below the Curie temperature. When the average value of the Higgs field is zero, it is postulated that the masses of all the elementary particles have the same value, namely zero. Under different conditions, this field develops a nonzero average value (i.e., a nonzero order parameter). It is this nonzero average value of the Higgs field that is believed to provide different masses for elementary particles. The "Higgs boson" is a particle related to the Higgs field. Evidence for this particle is being sought through the use of the high-energy accelerator at the LHC. If found, this experimental evidence will provide confirmation of Higgs' theory.

8.3 SOLUTIONS IN SEARCH OF PROBLEMS

In the normal sequence of problem solving, one starts with a problem and then tries to work out its solution through direct calculation. In some situations, however, this sequence is reversed: A general class of solutions might be known in advance, and it is applied to a specific problem. In this situation, we encounter "solutions in search of problems."

Some algorithms generate a set of expressions *guaranteed* to be solutions to a certain class of problems. If we have a problem that belongs to this class, then we might be able to find a solution that has already been generated for us. This type of indirect workflow is used in the study of some two-dimensional electrostatics problems, as illustrated next.

Usually, we deal with problems in three spatial dimensions (x, y, z). The class of problems considered here, however, is independent of the third spatial variable z. Consequently, the solution to these problems can be expressed solely in terms of the other two spatial coordinates x and y, and can be analyzed in two dimensions, e.g., in the xy-plane. Even if the problem has weak dependence on z, a two-dimensional analysis might provide a good approximation to the exact solution:

$$\varphi(x, y, z) \approx \varphi(x, y). \tag{8.42}$$

Consider a bar that is oriented in the z-direction and has a uniform electric charge per unit length. If the bar was *infinitely* long, then by translational symmetry along the z-direction, the mathematical expressions for the electrostatic potential φ, electric field vector $\vec{E}$, and related quantities cannot have any z-dependence, and the problem can be analyzed in two dimensions. Although real bars are finite in length, the two-dimensional model often provides a good approximation. This is particularly true when the radial distance from the point of interest to the bar is much less than the total length of the bar, and the z-coordinate of the point corresponds to the central portion of the bar.

Two dimensional problems are of special interest in the field of electrostatics because there is an algorithm that generates solutions for the electrostatic potential φ. Consider a general function $\varphi(x, y)$ that satisfies Laplace's equation in two dimensions:

$$\nabla^2 \varphi \equiv \frac{\partial^2 \varphi}{\partial x^2} + \frac{\partial^2 \varphi}{\partial y^2} = 0. \tag{8.43}$$

A function that satisfies Laplace's equation is said to be "harmonic." Equation 8.43 arises in a variety of physical situations. One of these is the study of electrostatics, where the scalar potential $\varphi(x, y)$ satisfies Equation 8.43 in any region where there is zero net electric charge density (i.e., zero charge per unit volume). Equation 8.43 also arises in the mathematical analysis of steady-state heat flow. There, the function φ describes the spatial distribution of temperature in a region where there are no heat sources.

Given the scalar potential φ, the electric field vector $\vec{E}$ is calculated from its gradient with the relationship

$$\vec{E}(x, y) = -\vec{\nabla}\varphi \equiv \left(-\frac{\partial\varphi}{\partial x}, -\frac{\partial\varphi}{\partial y} \right). \tag{8.44}$$

Because $\varphi(x, y)$ does not depend on the variable z, the component of the electric field along that direction is zero, i.e., $E_z = 0$.

We want to solve Laplace's equation for $\varphi(x, y)$ outside some arrangement of the conducting surfaces. The electrostatic potential $\varphi(x, y)$ must be a constant on any conducting surface so that no current flows in the absence of an external source, such as a battery. That boundary condition, along with Equation 8.43, defines the two-dimensional boundary value problem:

$$\nabla^2\varphi = 0 \text{ (outside the conductor)},$$
$$\varphi(x, y) = \text{const (on the surface of any conductor)}. \tag{8.45}$$

The solution of $\varphi(x, y)$ outside the conductor depends on the specific shape of conducting surface(s) and their specified voltages. Except for a few simple shapes, there is no direct, analytic way to solve the boundary value problem given in Equation 8.45. Using the tools of complex analysis, however, infinitely many solutions can be generated indirectly using the properties of Laplace's equation.

Consider the function $f(Z)$, where $Z = x + iy$ is a complex variable and the values of the function f are also complex. (We denote the complex variable by uppercase Z here to avoid possible confusion with the spatial coordinate z.) The complex function $f(Z)$ can be decomposed into two real functions,

$$f(Z) = u(x, y) + iv(x, y), \tag{8.46}$$

or equivalently

$$u(x, y) = \text{Re}(f(Z)),$$
$$v(x, y) = \text{Im}(f(Z)). \tag{8.47}$$

A remarkable theorem, which is proved in standard texts on complex analysis, states that provided the derivative $f'(Z)$ exists within a region R (i.e., $f(Z)$ is "analytic" in R), the Cauchy–Riemann equations

$$\frac{\partial u}{\partial x} = \frac{\partial v}{\partial y} \tag{8.48}$$

and

$$\frac{\partial u}{\partial y} = -\frac{\partial v}{\partial x} \tag{8.49}$$

are satisfied in that region. The proof is omitted here, but the main idea is as follows. Provided that the derivative $f'(Z)$ exists at some point in the complex plane, then we are free to calculate it either by taking the limit as $\Delta x \to 0$ or by taking the limit as $\Delta y \to 0$. The two results must be equal, and this leads to the Cauchy–Riemann equations.

Because the order of partial differentiation can be interchanged, e.g.,

$$\frac{\partial^2 u}{\partial x \partial y} = \frac{\partial^2 u}{\partial y \partial x}, \tag{8.50}$$

a further result of Equations 8.48–8.50 is that $u(x, y)$ and $v(x, y)$ both satisfy Laplace's equation:

$$\frac{\partial^2 u}{\partial x^2} + \frac{\partial^2 u}{\partial y^2} = 0 \tag{8.51}$$

and

$$\frac{\partial^2 v}{\partial x^2} + \frac{\partial^2 v}{\partial y^2} = 0 \tag{8.52}$$

Consequently, *any* differentiable, complex function $f(Z)$ produces two real functions u and v. Both are guaranteed to be solutions to the two-dimensional Laplace's equation.

EXAMPLE 8.5

Consider the quadratic function $f(Z) = Z^2$.

(a) Using this function, find the functions u and v corresponding to its real and imaginary parts.
(b) Show that both u and v satisfy Laplace's equation.
(c) Consider the sets of contour lines

$$u(x, y) = C_1, \tag{8.53}$$

where C_1 is a constant and the analogous expression for v is

$$v(x, y) = C_2. \tag{8.54}$$

Make contour plots of constant values of $u(x, y) = C_1$ for $C_1 = -3, -2, -1,$ $0, +1, +2, +3$ and $v(x, y) = C_2$ for $C_2 = -3, -2, -1, 0, +1, +2, +3$.
(d) Identify the electrostatic potential to be $\varphi(x,y) = Ku(x, y)$, where K is a constant that has units of volts per square meter. Find an expression for the electric field vector $\vec{E}(x, y)$.

ANSWER

(a) The Cauchy–Riemann equations for $f(Z) = Z^2$ hold because its derivative $F'(Z) = 2Z$ exists everywhere in the complex plane. We have

$$f(Z) = Z^2 = (x + iy)^2 = x^2 - y^2 + 2ixy. \tag{8.55}$$

Applying Equation 8.47 to Equation 8.55 yields

$$u(x, y) = x^2 - y^2 \tag{8.56}$$

and

$$v(x, y) = xy. \tag{8.57}$$

(b) The Laplacian of $u(x, y) = x^2 - y^2$ is

$$\nabla^2 u(x, y) = \left(\frac{\partial^2}{\partial x^2} + \frac{\partial^2}{\partial y^2} \right)(x^2 - y^2) = 2 - 2 = 0. \tag{8.58}$$

The Laplacian of v is also zero,

$$\left(\frac{\partial^2}{\partial x^2} + \frac{\partial^2}{\partial y^2} \right)(xy) = \left(\frac{\partial y}{\partial x} + \frac{\partial x}{\partial y} \right) = 0 + 0 = 0. \tag{8.59}$$

Equations 8.58 and 8.59 verify that both u and v are harmonic functions, as expected.
(c) Figure 8.7 shows the contours of u and v. The contours have hyperbolic shapes, except when the constants are equal to 0, in which case the contours are straight lines. If $u(x, y)$ was proportional to an electrostatic potential, $\varphi(x, y) = Ku(x, y)$, then Equation 8.53 describes lines of equipotential, i.e., $\varphi =$ constant.

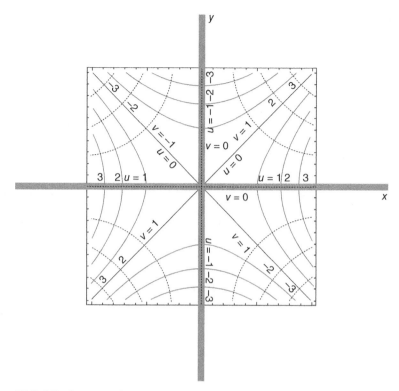

FIG. 8.7. Contour plot for the functions u and v derived in Example 8.5.

(d) From Equation 8.44, the electric field for the electrostatic potential considered in part (c) is

$$\vec{E}(x, y) = -\vec{\nabla}\varphi = 2K(-x, y). \qquad (8.60)$$

The pair of functions u and v generated by $f(Z)$ have another interesting property. As discussed in texts on vector calculus, the gradient of $u(x, y)$

$$\vec{\nabla}u = \left(\frac{\partial u}{\partial x}, \frac{\partial u}{\partial y}\right) \qquad (8.61)$$

is a vector quantity that points in the direction of greatest change of $u(x, y)$ so that its direction is perpendicular to the contour lines representing $u(x, y) =$ constant. The dot product of the gradients of u and v is

$$\vec{\nabla}u \cdot \vec{\nabla}v = \frac{\partial u}{\partial x}\frac{\partial v}{\partial x} + \frac{\partial u}{\partial y}\frac{\partial v}{\partial y}. \qquad (8.62)$$

Combining Equation 8.62 with Equations 8.48 and 8.49 yields

$$\vec{\nabla}u\cdot\vec{\nabla}v = 0. \tag{8.63}$$

Equation 8.63 implies that the contours of u and v described in Equations 8.53 and 8.54 must be perpendicular to each other at any point where they intersect. This property is apparent for the specific case shown in Fig. 8.7. It is also true in general for the pair of functions associated with the real and imaginary parts of *any* analytic function of Z. To verify the result, first choose any point (x, y) and draw a line that is perpendicular to the contours of the function u passing through that point. Next, draw a second line that is perpendicular to the contour of v passing through the same point. Equation 8.63 implies that the two lines must be perpendicular to each other. Because the analysis of the problem is restricted to a plane, Equation 8.62 requires that the contours of u and v must be perpendicular to each other at any point of intersection.

This orthogonality property provides information about the electric field vector $\vec{E}$ given in Equation 8.44. Suppose the function u corresponds to the electrostatic potential φ. Then, from Equations 8.44 to 8.63, the contour lines of its paired function v provide the orientation (but not the sign) of electric field vector at each point in space. The plot of this orientation is sometimes called the *stream function*. The sign, i.e., the specific direction of the electric field along the stream function, can be determined from voltage of the conductors that define the boundary conditions. Electric field lines originate from conductors at higher voltage (i.e., carrying more positive charge) and terminate on conductors at lower voltage (i.e., carrying more negative charge).

We can obtain qualitative information about the strength of the electric field from the stream function by examining how densely the contours are packed together. The regions in which the contours are more densely packed correspond to regions of stronger electric field, i.e., the magnitude of $\vec{E}$ is greater. The function v itself does not provide the functional form of the electric field vector, but that information is readily available by calculating the gradient of u using Equation 8.44. Finally, if it better suits the boundary conditions provided by the specified conducting surfaces, we are free to choose v instead of u to be proportional to the electrostatic potential. With that choice of roles, the contour lines of u provide the orientation of the electric field.

Finding the solution that corresponds to the specific boundary conditions given for φ can be difficult. The possible choices for the generating function $f(Z)$ are in principle endless, which, depending on one's goal, could be a help or a hindrance. Today, many electrostatics problems are solved directly using numerical methods, but historically, indirect methods like the use of generating functions $f(Z)$ were very useful. These indirect methods can be used to explain important qualitative features of electrostatics. For example, the experimental observation that electric fields tend to be strong near sharp points or protuberances from conductors (hence, lightning rods have sharp points) can be readily understood by visualizing sets of u and v contour consistent with conducting surfaces.

If an indirect method generates a particularly interesting or useful solution to an electrostatics problem, we can fabricate conductors (as boundary conditions) into the specific shape and fix their voltages to the desired quantities using batteries or DC power supplies. This is done in the construction of an electrostatic quadrupole lens, which is made from four sheets of metal. Referring to Fig. 8.7, consider the four contours with $|u| = 1$. We hold the two contours labeled $u = 1$ at equal positive voltage $+V_0$ and the two contours labeled $u = -1$ at negative voltage $-V_0$. According to Equation 8.60, the resulting electric field is given by $\bar{E}(x, y) = 2V_0(-x, y)$ or $E_x = -2V_0 x$. Therefore, the x-component of the electric field accelerates a positively charged particle toward the line $x = 0$. This provides a focusing effect analogous to that of an optical lens. Further discussion of topics related to Example 8.5 is provided in volume 2 of Feynman et al. (2005).

8.4   LEARNING FROM REMARKABLE RESULTS

In Chapter 5, we discussed a differential equation that is used as a simple model for population growth,

$$\frac{dp(t)}{dt} = Kp(t) - Ap^2(t), \tag{8.64}$$

and solved it with direct integration (see Eq. 5.18). When the constants K and A are positive, the first term on the right-hand side of Equation 8.64 represents population growth, and the second term represents population decline due to scarcity of food, overcrowding, and other constraints. Over time, these two competing terms balance each other, and the solution to Equation 8.64 reaches a steady state. Recall that the steady-state solution p_{ss} to the differential equation is found by setting the derivative equal to zero, so that

$$p_{ss} = \frac{K}{A}. \tag{8.65}$$

Here, we revisit this topic by analyzing the *logistic difference equation*, which is related to Equation 8.64, but is distinct from that differential equation. A difference equation serves as a more appropriate tool than a differential equation to model situations where events occur at discretely separated time intervals, rather than evolving over time in a continuous fashion. For example, suppose that $p(t)$ represents the population of wolves in a habitat. Assuming that wolf pups are born only in the spring of the year, it is reasonable to replace the continuously varying function $p(t)$ with the discrete variable p_n, where $n = 0, 1, 2, 3 \ldots$ years.

The logistic difference equation is given by the nonlinear recursive relation

$$x_{n+1} = r\left(x_n - x_n^2\right), \tag{8.66}$$

where r is a positive constant, typically taken in the range $0 < r < 4$. We restrict x_n to the range $0 < x_n < 1$ to ensure that the right-hand side of Equation 8.66 also remains positive. We can think of $x_n = 0$ as representing the extinction of the population and $x_n = 1$ as representing the maximum population that the habitat could support.

Beginning in the 1960s, it was observed that nonlinear difference equations such as Equation 8.66 can give rise to solutions that are far more intricate than the single steady-state value $p_{ss} = K/A$, which is obtained from Equation 8.64. The strange and interesting behavior that emerges cannot be anticipated from the solutions of Equation 8.64 alone. This realization was an important advancement in the development of the chaos theory, which is characterized by complicated behavior emerging from simple equations. The solutions to Equation 8.66 illustrate properties such as bifurcations and the butterfly effect, and a remarkable result known as universality.

EXAMPLE 8.6

(a) Show how Equation 8.66 is related to Equation 8.64.
(b) Discuss some of the main features that emerge for the solution x_n in Equation 8.66 as r is increased from 0 to 4.

ANSWER

(a) Dimensional checks and the use of non-SI units introduced in Chapter 1 can guide us when establishing the relationship between the differential and difference equations. The variable x_n in Equation 8.66 must be dimensionless because a squared term is subtracted from a linear term. Because x_n is dimensionless, we can infer from Equation 8.66 that the positive constant r is also dimensionless.

The variables in Equation 8.64, however, do carry dimensions. The variable t has dimensions of time, which we will choose to measure in units of years. It is convenient to introduce a non-SI unit for the population p, which we will call the "wolf." Let us introduce a positive conversion factor w, also measured in wolves. Then, in order to find the correspondence between Equations 8.64 and 8.66, we define a dimensionless (but continuously varying) quantity:

$$x(t) = \frac{p(t)}{w}. \tag{8.67}$$

Then, the units of the relevant variables in Equation 8.64 are

$$[p(t)] = \text{wolf},$$
$$[t] = \text{year},$$
$$[K] = \text{year}^{-1}, \tag{8.68}$$
$$[A] = \text{wolf}^{-1} \times \text{year}^{-1}.$$

and

$$[w] = \text{wolf},$$
$$[x(t)] = 1 \text{ (i.e., dimensionless).} \tag{8.69}$$

Using Equation 8.67 to eliminate $p(t)$ from Equation 8.64 yields

$$\frac{dx(t)}{dt} = Kx(t) - wAx^2(t) \tag{8.70}$$

or, in terms of differentials,

$$dx = Kx(t)dt - wAx^2(t)dt \tag{8.71}$$

Now we can consider the year-to-year variation in the population by letting $dt \to \Delta t$ and $x(t) \to x_n$. In this discrete form, Equation 8.71 becomes

$$\Delta x = x_{n+1} - x_n = Kx_n\Delta t - wAx_n^2\Delta t. \tag{8.72}$$

Note that each term of Equation 8.72 is dimensionless. During any year n, the population of wolves is obtained from the dimensionless variable x_n by applying the scale factor w:

$$p_n = wx_n. \tag{8.73}$$

Comparing Equations 8.66 and 8.72, we can infer that

$$r = K\Delta t + 1,$$
$$r = A\Delta tw. \tag{8.74}$$

Eliminating Δt from Equation 8.74 and solving for w yields

$$w = \frac{K}{A} \times \left(\frac{r}{r-1}\right), \tag{8.75}$$

so that the population of wolves in year n is given by

$$p_n = \frac{K}{A} \times \left(\frac{r}{r-1}\right) x_n. \tag{8.76}$$

(b) Consider the solution x_n to the logistic difference equation for various values of r. Equation 8.66 does not have analytic, closed-form solutions, except for a few select values of r. For any value of r, however, it is relatively easy to iterate Equation 8.66 with a computer or a programmable calculator. For example, we can take a starting value such as $x_0 = 0.5$ and then iterate $x_1 = r(0.5 - 0.5^2)$, and so on. We need to calculate a sufficient number of iterations so that the transients decay away and the long-term behavior emerges.

A graphical procedure known as a *web map* can also be very helpful, and three sample cases are illustrated in Fig. 8.8a–c. (Note the similarity between the web map and the Jones plot discussed in Chapter 4.) Each web map displays a plot of x_{n+1} versus x_n. Figure 8.8a–c illustrates the effect of changing the value of r on the solution for large values of n. When $0 < r < 1$, x_n tends to zero after a number of iterations, so that the population dies out. This is expected because from Equation 8.75, we see that $K \propto r - 1$, so that K becomes negative for $r < 1$, and the solutions to Equation 8.64 contain exponential decay.

For $1 < r < 3$, the numerical simulations readily show that

$$\lim_{n \to \infty} x_n = \frac{r-1}{r}, \quad 1 < r < 3. \tag{8.77}$$

The convergence of x_n to 0 or $(r-1)/r$ is not at all surprising, because the intersection points of the line and parabola in Fig. 8.8b occur at $x = 0$ and $x = (r-1)/r$. For $1 < r < 2$, the convergence in Equation 8.77 is rapid, while for $2 < r < 3$, there is some oscillation before the value $(r-1)/r$ is reached. Combining Equations 8.76 and 8.77, we see that

$$\lim_{n \to \infty} p_n = \frac{K}{A}, \quad 1 < r < 3, \tag{8.78}$$

which agrees with the steady-state solution of the differential equation, given in Equation 8.65. Up to this point, everything seems to be fitting together very neatly.

The remarkable and unexpected results begin to emerge for values of $r = 3$. For r slightly greater than 3, the iterated values of x_n oscillate between two values above and below $(r-1)/r$, and no single steady-state population is reached. For example, we find that (to three significant figures) when $r = 3.1$, $(r-1)/r \approx 0.677$, and x_n oscillates between 0.558 and 0.765. The period of oscillation is 2 years. If the pattern for x_n repeats once, it will continue to repeat forever. This is because, according to Equation 8.66, the population in year n completely determines the population for the following year, assuming that r is fixed. As r is further increased beyond

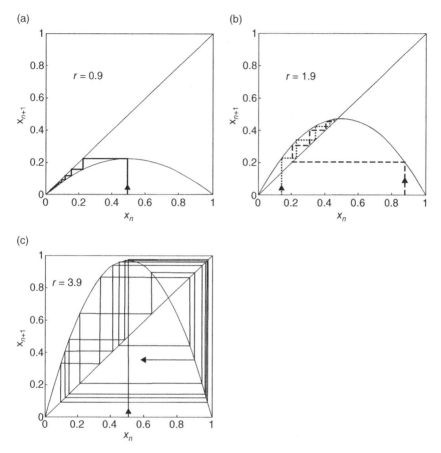

FIG. 8.8. The logistic difference equation, Equation 8.66, can be solved with a graphic procedure known as a web map. (a) When the parameter $r = 0.9$, x_n converges to zero regardless of the starting value x_0. (b) When $r = 1.9$, x_n converges to $(r-1)/r = 0.474$ regardless of the starting value x_0. (c) When $r = 3.9$, chaotic behavior is observed. Unlike (a) and (b), there is extreme sensitivity to the starting value.

the threshold $r = 3$, the gap between the two x_n values increases. We say that x_n has a "bifurcation" at $r = 3$, and we will label this bifurcation point r_2.

For further increases in the parameter r, the results get even stranger. There is another bifurcation at $r_4 = 1 + \sqrt{6} = 3.449490$, and the oscillations go through a 4-year cycle. For example, when $r = 3.5$, x_n cycles through the values 0.383, 0.827, 0.501, and 0.875. At $r_8 = 3.544090$, the cycle changes from a 4-year pattern to an 8-year pattern, then to a 16-year pattern when $r_{16} = 3.564407$. The values of r at which the pattern bifurcates get closer and closer together until the critical value of $r_\infty = 3.569945672$ is reached, which is called an *accumulation point*. Beyond the accumulation point, the solution to Equation 8.66 has a chaotic region where the pattern never repeats.

In the chaotic region, the value of x_n displays extreme sensitivity to the starting value x_0. For example, setting $r = 3.9$ with $x_0 = 0.5000$, we find that after 30 iterations, $x_{30} = 0.9728$. If, however, $x_0 = 0.5001$, a tiny change in the initial condition, then $x_{30} = 0.2845$, which is a considerable difference. This is sometimes referred to as the "butterfly effect." The nonlinear equations that model weather also have solutions that display chaos. Taken to the extreme, a tiny change in the initial wind conditions caused by the fluttering of a butterfly's wings in one part of the world could have a profound effect on the weather across the globe a few weeks later. Because of the butterfly effect, the prospect of accurately making detailed weather forecasts far in advance (e.g., 30 days) seems hopeless, regardless of the sophistication of the mathematical models, quality of the input data, and the amount of the computer power. This is one way in which chaos theory has fundamentally changed our perception of the world.

During the 1970s, Mitchell Feigenbaum, an American mathematical physicist, further extended these remarkable results. He examined the same logistics difference equation as described in Example 8.6 and observed the patterns we described. He determined that the ratio of the gaps between successive threshold values given by the sequence

$$\frac{r_4 - r_2}{r_8 - r_4}, \frac{r_8 - r_4}{r_{16} - r_8}, \frac{r_{16} - r_8}{r_{32} - r_{16}}, \ldots \tag{8.79}$$

gets smaller in a systematic progression. The ratio converges to what is called the first Feigenbaum constant $\delta = 4.6692\ldots$.

Feigenbaum also observed that the results do not depend on the exact form of Equation 8.66. Other nonlinear difference equations with a "hump" work equally as well. For example, the solutions of $x_{n+1} = r \sin(\pi x_n)$ display the same bifurcations in a progression toward chaotic behavior. Although the specific values of r_2, r_4, r_8 differ from those obtained with the logistic equation, the ratios in Equation 8.79 also converge to exactly the same value δ. Several examples are provided in Sprott (2003; a very complete and accessible presentation of many topics in chaos theory is provided in this book).

This broad connection between the solutions of seemingly unrelated equations is referred to as "universality." Universality was discussed in connection with scale invariance and phase transitions in Chapter 6. In the context of the logistic equation, very few nonlinear problems can be solved analytically. The principle of universality allows what is learned from one problem to be applied to another, so it is extremely useful. This is somewhat like the electromechanical analogy described between the RLC electrical circuit and damped oscillator described in Chapter 7, only on a much grander scale.

The final worked example problem in this book deals with the early history of the universe. In 1965, Arno Penzias and Robert Wilson at the Bell Laboratories made a discovery that earned a Nobel Prize in physics and ushered in a new era for the study of cosmology. They discovered what has become known as the "3-degree" microwave background radiation.

The existence of the background radiation was first predicted by the American physicists Ralph Alpher and Robert Herman, who used estimates of the abundance of nuclei heavier than hydrogen to arrive at a density and temperature for the early universe (Alpher and Herman 1948). Today, astronomers can detect this steady background of electromagnetic radiation coming from any direction in space. They find that this radiation has a spectrum characteristic of a blackbody at a temperature of 2.7 K (see Eq. 3.14). The next example uses scaling arguments to enable us to look back in time approximately 15,000,000,000 years to the early universe. The solution draws on a principle describing chemical reactions, called the law of mass action, as well as principles from the field of cosmology. Even if the reader is not familiar with these particular principles, the resulting mathematics is quite straightforward.

EXAMPLE 8.7

Hydrogen is the most abundant chemical element in the universe. Perhaps, up to 100,000 years after the big bang, hydrogen existed mainly in an ionized form consisting of free negatively charged electrons and positively charged protons. It is believed that the average temperature of the early universe was much higher than it is today. Then, after the universe expanded and cooled sufficiently, a great number of protons and electrons combined to form electrically neutral hydrogen atoms. Before that transition, the universe was filled with electromagnetic radiation (i.e., photons, mostly in the X-ray energy range), which was in thermal equilibrium with matter at the prevailing temperature of the universe. At the transition, two things happened. First, many more photons were created as protons and electrons combined with each other. Second, a "freeze-out" of electromagnetic radiation occurred, after which it ceased to be in thermal equilibrium with matter. This freeze-out occurred because neutral hydrogen atoms are essentially transparent to X-rays, while free electrons scatter X-ray photons very efficiently. The 3-degree background microwave radiation observed today is believed to be the remnants of the electromagnetic radiation dating from the era of the freeze-out.

Current theories of cosmology assume that the universe is expanding and can be characterized by a radius R, which grows over time. We can apply the concepts of scaling introduced in Chapter 6 and use the ratio between the current radius and the radius around the time of the radiation freeze-out to study the history of the universe.

(a) Use a principle from cosmology to set up a scaling relationship between the wavelength of the background radiation and the radius of the universe.

(b) Using the concept of blackbody radiation introduced in Chapter 3, find a relationship between the peak of the distribution of observed wavelengths in the background radiation and the scale factor determined in part (a). Use this relationship to infer information about the average temperature of the universe around the time of the radiation freeze-out.

(c) Use a scaling argument to estimate the density (number of particles *per unit volume*) of protons and hydrogen atoms at the time of the radiation freeze-out, assuming the current value for the density of hydrogen is known.

(d) Write a reaction describing the equilibrium between ionization and recombination of hydrogen before the radiation freeze-out occurred.

(e) Using the *law of mass action* from chemistry, develop an expression that describes the densities for the reaction described in part (d).

(f) Estimate the temperature T_{fo} at radiation freeze-out from the equations developed in the previous steps. Estimate the extent of the expansion of the universe since the radiation freeze-out.

ANSWER

(a) Under assumptions of cosmology, the wavelength λ of electromagnetic radiation increases as the universe expands. The wavelength scales linearly with the change in radius of the universe according the relationship

$$\lambda_{curr} = \lambda_{fo} \frac{R_{curr}}{R_{fo}} \equiv \lambda_{fo} S_R, \tag{8.80}$$

where the subscript "curr" indicates current values, and "fo" indicates values at the time of the freeze-out. Equation 8.80 represents the *cosmological redshift*, which is described further in Weinberg (1993; a very readable account of the early universe is provided in this book). The dimensionless scale factor S_R is much like the scale factors introduced in Example 6.4 for the discussion of the scale model of the barge. The scale factor $S_R > 1$ in Equation 8.80 because the universe has expanded since the radiation freeze-out.

(b) We assume that prior to freeze-out, the radiation was in thermal equilibrium with the electron–proton mixture. Consequently, its spectrum (i.e., its distribution of frequencies or wavelengths) can be characterized by Planck's formula that was discussed in Chapter 3; see Equation 3.14.

Exercise 8.5 outlines how Planck's formula can be reexpressed by replacing the frequency variable f as in Equation 3.14 with the wavelength variable λ. The two are related by $\lambda = cf$, where c is the speed of light.

The shape of the spectrum of blackbody radiation and the location of its maximum λ_{peak} depend on the absolute temperature T. As derived in exercise 8.5,

$$\lambda_{peak} = \frac{b}{T}, \tag{8.81}$$

where the constant $b \approx 2.898 \times 10^{-3}\,K{\cdot}m$. Equation 8.81 is known as Wien's displacement law. From Equation 8.81, we can deduce that

$$\frac{\lambda_{peak,curr}}{\lambda_{peak,fo}} = \frac{T_{fo}}{T_{curr}}. \tag{8.82}$$

From Equations 8.80 and 8.82, we obtain

$$T_{curr} = \frac{T_{fo}}{S_R}. \tag{8.83}$$

Equation 8.83 tells us that the currently observed background radiation, which has a spectrum characteristic of a blackbody at the temperature $T_{curr} = 2.7\,K$, actually was associated with a much higher temperature at the time of the radiation freeze-out.

(c) Although elemental helium was created before the freeze-out, and heavier elements have been synthesized in stars since then, hydrogen has always been the most prevalent element in the universe. As an approximation, we will assume that the total number of protons plus hydrogen atoms has not changed appreciably (relative to the total number) since the radiation freeze-out. This assumes that the number of protons has been conserved, i.e., protons do not decay into other types of elementary particles. Under these assumptions, the density of protons and hydrogen atoms n_{fo} can be estimated by scaling from the current density n_{curr}:

$$n_{fo} = n_{curr}S_R^3. \tag{8.84}$$

Through Equations 8.83 and 8.84, the expansion scale factor S_R links both the temperature and density at the time of freeze-out to their currently observed values.

(d) The reaction we consider is the combination of a proton p and an electron e to form a neutral hydrogen atom H. It requires $\Delta E = 13.6\,eV$ (i.e., $2.18 \times 10^{-18}\,J$) of energy for the reverse reaction, i.e., to ionize one hydrogen atom. When a proton and electron combine at rest, this energy is released in the form of an X-ray photon γ with energy ΔE. The two-way reaction that describes the ionization and recombination is

$$p + e \rightleftarrows H + \gamma. \tag{8.85}$$

Let the fraction of hydrogen in the ionized state be

$$f = \frac{n_\mathrm{p}}{n_\mathrm{p} + n_\mathrm{H}}, \tag{8.86}$$

which ranges from 1 to 0. We expect that at the higher temperatures encountered in the earliest universe, the fraction f was close to 1, and its value has decreased over time with cooling. It is convenient to work with the parameter f because it is expected to fall precipitously during the radiation freeze-out as electrically neutral hydrogen is formed.

(e) We can apply an equation from physical chemistry called the *law of mass action* that allows us to calculate the relative number densities n (i.e., number of particles per unit volume) of constituent species for an equilibrium reaction such as Equation 8.85. In this context, thermal equilibrium means that the rates for the reactions described in Equation 8.85 are equal in both directions. This equality ensures that, over an intermediate timescale, the average number of each type of particle is constant. This is a reasonable assumption because of the fast timescale of an individual reaction and the enormous timescale of the expansion of the universe.

In previous chapters, we discussed the chemical potential. The law of mass action can be developed using these quantities because the conservation of energy relates the chemical potentials of protons and electrons to those of hydrogen atoms and photons. The details are omitted here, but an analysis that uses the chemical potentials of ideal gases provides the following law of mass action for the reaction represented by Equation 8.85:

$$\frac{n_\mathrm{e} n_\mathrm{p}}{n_\mathrm{H}} = \frac{n_{Q\mathrm{e}}(T) n_{Q\mathrm{p}}(T)}{n_{Q\mathrm{H}}(T)} \exp\left(\frac{-\Delta E}{kT}\right). \tag{8.87}$$

This expression is also known as a Saha equation, in honor of the Indian astrophysicist Megh Nad Saha who published his work in the 1920s. Each of the factors $n_Q(T)$ in Equation 8.87 has dimensions of length^{-3} and can be expressed as a function of the mass m, temperature T, Planck's constant h, and Boltzmann's constant k:

$$n_Q = \left(\frac{2\pi m k T}{h^2}\right)^{3/2}. \tag{8.88}$$

Those who have studied statistical mechanics or physical chemistry might recognize the n_Q factors as the inverse cube of the so-called "thermal wavelength."

To further simplify Equation 8.87, we will also assume that the universe has no net charge. If it had, then electrostatic forces would dominate over gravitational forces. Astronomers do not observe those effects, so the assumption is reasonable. Furthermore, we will assume that the positive charge is provided by protons and the negative charge by electrons so that their average densities (in terms of the number of particles per volume) are equal to each other:

$$n_p = n_e. \tag{8.89}$$

Combining Equations 8.87–8.89 and using $m_p \approx m_H$ yields a scaling relationship for density and temperature before radiation freeze-out:

$$\frac{n_p^2}{n_H} = \left(\frac{2\pi m_e kT}{h^2}\right)^{3/2} \exp\left(\frac{-\Delta E}{kT}\right). \tag{8.90}$$

Using the definition of the fraction of ionized hydrogen f in Equation 8.86, we can express the ratio appearing in Equation 8.90 as

$$\frac{n_p^2}{n_H} = \frac{f^2}{(1-f)}(n_p + n_H). \tag{8.91}$$

Then, Equation 8.90 can be expressed as

$$\frac{f^2}{(1-f)} = \frac{1}{(n_p + n_H)} \times \left(\frac{2\pi m_e kT}{h^2}\right)^{3/2} \times \exp\left(\frac{-\Delta E}{kT}\right). \tag{8.92}$$

The densities and temperatures appearing in Equation 8.92 correspond to those existing around the time of the freeze-out.

(f) Using the scale factor S_R, we can now relate the densities and temperatures around the time of the freeze-out to the values observed today. The value of the temperature is given in Equation 8.83. Assuming that $n_p + n_H = n_{H,curr}$ and applying Equations 8.84 and 8.92 lead to the following relationship between f and S_R:

$$\frac{f^2}{(1-f)} = \frac{1}{S_R^3 n_{H,curr}} \times \left(\frac{2\pi m_e k S_R T_{curr}}{h^2}\right)^{3/2} \times \exp\left(\frac{-\Delta E}{k S_R T_{curr}}\right) \equiv G(S_R). \tag{8.93}$$

Most of the quantities in Equation 8.93 have known values, including h, k, the mass m_e of the electron, $\Delta E = 13.6\,eV$, and $T_{curr} = 2.7\,K$. Also, based on astronomical observations, it is believed that the current, average density of hydrogen atoms per unit volume in the universe is on the order of $n_{\text{H,curr}} \approx 2 \times 10^{-7}\,\text{m}^{-3}$. This means the only unknown quantities in Equation 8.93 are f and S_R, which are both dimensionless. Equation 8.93 reduces to the manageable form

$$f^2 + Gf - G = 0. \tag{8.94}$$

with

$$G(S_R) = A S_R^{-3/2} e^{-(B/S_R)}. \tag{8.95}$$

The dimensionless constants are $A \approx 5 \times 10^{28}$ and $B \approx 5.85 \times 10^4$. Because $f > 0$, the physically relevant solution Equation 8.94 is

$$f = \frac{\sqrt{G^2 + 4G} - G}{2}. \tag{8.96}$$

If we knew the value of f at the time of the freeze-out, then we could use Equation 8.93 to find S_R. Even without knowledge of the precise value of f, we can substitute different values of S_R into Equations 8.95 and 8.96 and then plot the corresponding values of f. The results are shown in Fig. 8.9.

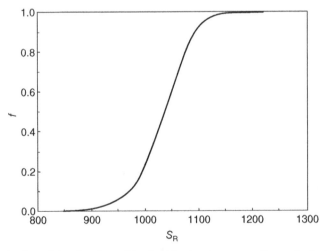

FIG. 8.9. A plot of the fraction of ionized hydrogen versus the scale factor S_R for the expansion of the universe, as modeled by Equation 8.93. Earlier times correspond to larger values of S_R. The radiation freeze-out occurred when there was a sudden drop in the value of f.

Working with these equations provides answers that are consistent with our expectations. As the universe expands, i.e., S_R increases, the temperature falls, and the fraction of ionized hydrogen f falls. As illustrated in Fig. 8.9, the fraction f plummets in the range of $S_R \approx 1000$–1100. From Equation 8.83, this corresponds to a temperature of $T \approx 2700$–3000 K. We can assume that the freeze-out occurred somewhere in this range. Thus, even without knowing the precise value of f at the time of the radiation freeze out, we can still make reasonable estimates for T_{fo} and S_R. Using this information and applying other results from cosmology about the expansion of the universe allow us to date the freeze-out at roughly 10^5 years after the big bang.

In summary, the steps outlined in Example 8.7 have allowed us to account for the source of the observed microwave background radiation and its observed distribution of wavelengths. This was done by manipulating relatively simple equations borrowed from the fields of cosmology and chemistry. In addition, this analysis allowed us to estimate the size of the universe at the time when radiation ceased to be in thermal equilibrium with matter, or the "freeze-out." Introducing a scale factor S_R for the radius of the universe at different times provided a link between the early-time density of hydrogen and the value observed presently, and also linked the temperature at the time of the freeze-out to the temperature corresponding to the observed microwave background radiation.

EXERCISES

(8.1) Show that a *two-dimensional* corner reflection as shown in Fig. 8.1a has the desired property by directly applying Equation 8.1 and tracing the reflected rays geometrically. Compare the amount of work that is required with the amount of work to apply the vector component method described in Example 8.1. Is it feasible to use geometric ray tracing and Equation 8.1 to demonstrate the properties of a three-dimensional corner reflector, i.e., a corner cube?

(8.2)

(a) Use Equations 8.9, 8.10, and 8.19 to show that $s^2 - y^2 = 2T_H y / w$ for a catenary. Interpret the result of part (a) under the following conditions: (b) $w \to \infty$, (c) $T_H \to 0$, and (d) $s = y$.

(8.3) Find the electrostatic potential φ outside of a grounded conducting sphere of radius R centered at the origin when a point charge q is placed at $x = a \, (a > R)$ along the x-axis using the following steps.

Assume that the value of the image charge is βq and its location is at $x = b$ (with $b < R$) along the x-axis. The surface of the sphere is described by the equation $x^2 + y^2 + z^2 = R^2$, or $y^2 + z^2 = R^2 - x^2$. Thus, the requirement that the sphere is grounded translates (in SI units) to the equation

$$0 = \frac{1}{4\pi\varepsilon_0}\left(\frac{q}{\sqrt{(x-a)^2 + (R^2 - x^2)}} + \frac{\beta q}{\sqrt{(x-b)^2 + (R^2 - x^2)}}\right). \tag{8.97}$$

(a) Solve for β and b in terms of a and R by simplifying Equation 8.97 and requiring that the coefficients of the remaining powers of x are zero. Show that this procedure yields $b = R^2/a$ and $\beta = -R/a$. So, for points outside the sphere,

$$\varphi(x, y, z) = \frac{q}{4\pi\varepsilon_0}\left(\frac{1}{\sqrt{(x-a)^2 + y^2 + z^2}} - \left(\frac{R}{a}\right)\frac{1}{\sqrt{\left(x - \frac{R^2}{a}\right)^2 + y^2 + z^2}}\right). \tag{8.98}$$

(b) Show that Equation 8.98 is dimensionally correct, and check that it makes sense for the special cases $R = 0$ and $R = a$.

(8.4) Consider the generalization of Equation 8.55 to a power-law function,

$$f(Z) = Z^a,$$

which reduces to the previous special case when $a = 2$. Introduce polar coordinates r and θ:

$$x = r\cos\theta,$$
$$y = r\sin\theta.$$

(a) Show that expressing Z in polar form yields,

$$Z = re^{i\theta},$$

so that

$$Z^a = r^a\left(e^{i\theta}\right)^a = r^a e^{i\theta a}$$

(b) Apply Euler's identity to show that

$$u(r, \theta) = r^a\cos(a\theta),$$
$$v(r, \theta) = r^a\sin(a\theta).$$

(c) Express u and v solely in terms of the Cartesian coordinates x and y and show that

$$u(x, y) = (x^2 + y^2)^{a/2} \cos\left(a \arctan\left(\frac{y}{x}\right)\right),$$

$$v(x, y) = (x^2 + y^2)^{a/2} \sin\left(a \arctan\left(\frac{y}{x}\right)\right).$$

(d) It is quite messy to verify that $u(x, y)$ and $v(x, y)$ satisfy Laplace's equation directly in Cartesian coordinates. Instead, it is more convenient to calculate the Laplacian of these functions using polar coordinates. Assuming that $r \neq 0$, the Laplacian is given by

$$\nabla^2 f = \frac{\partial^2 f(r, \theta)}{\partial r^2} + \frac{1}{r}\frac{\partial f(r, \theta)}{\partial r} + \frac{1}{r^2}\frac{\partial^2 f(r, \theta)}{\partial \theta^2}.$$

Use this form to verify that $\nabla^2 u(r, \theta) = 0$ and $\nabla^2 v(r, \theta) = 0$. This is an example of the simplification offered by a clever choice of coordinate system.

(8.5)

(a) Represent Planck's law in terms of the wavelength λ. Specifically, starting with Equation 3.14, calculate $u_{P,\lambda}(\lambda, T)$ (i.e., the energy density per unit volume per unit wavelength) using the relationship, or,

$$u_{P,\lambda}(\lambda, T) = \left|\frac{df}{d\lambda}\right| \times u_P(f, T) = \frac{c}{\lambda^2} u_P(f, T),$$

to obtain

$$u_{P,\lambda}(\lambda, T) = \frac{8\pi hc}{\lambda^5}\frac{1}{\exp\left(\dfrac{hc}{\lambda kT}\right) - 1}.$$

(b) Find the maximum of the expression in part (a). Differentiate $u_{P,\lambda}(\lambda, T)$ with respect to λ, and set the result equal to zero. Solve the resulting expression numerically to find the value of λ_{peak}, i.e., demonstrate Wien's displacement law given in Equation 8.81.

(8.6) Find the wavelength and frequency of electromagnetic radiation that correspond to:
(a) ΔE, the ionization energy of hydrogen (hint: use the relationship $hf = \Delta E$);
(b) $T = 2.7\,\mathrm{K}$ (hint: use the relationship $hf = kT$).

(8.7) Referring to Equation 8.96, use the binomial series to show:
  (a) $f \rightarrow 1$ as $G \rightarrow \infty$;
  (b) $f \rightarrow \sqrt{G}$ as $G \rightarrow 0+$.
  (c) Returning to Equation 8.93, verify that cooling (i.e., decreasing the temperature T) always causes a decrease in the value of f. One way to verify this result is to verify that the function $f^2/(1-f) = G$ monotonically increases as f increases from 0+ to 1, and G also monotonically increases as T increases.

(8.8) In Chapter 5, we discussed Fermat's principle of least time, which states that when light travels from point A to point B, it traverses the particular path that requires the least total elapsed time. In this chapter, we discussed the law of reflection, as expressed by Equation 8.1. Consider light emanating from a fixed point source A, reflecting from a planar mirror, and then traveling to a predetermined observation point B. Naturally, along each of the two segments, the ray of light travels along a straight line. (This fact itself is consistent with, and required by, Fermat's principle.)
  (a) Draw a diagram illustrating that there are many possible paths for the ray of light to travel from point A, reflect from the mirror, and then reach point B.
  (b) Of all the possible paths, show that Fermat's principle of least time requires that the actual path taken obeys the law of reflection. Hint: Show that this problem is essentially equivalent to Example 8.3. Then, use geometry, along with the fact that the triangle in Fig. 8.5 is isosceles, to argue that Equation 8.1 must be satisfied.

REFERENCES

Alpher RA and Herman R. 1948. On the relative abundance of the elements. *Physical Review* 74: 1577.

Feynman RP, Leighton RB, and Sands M. 2005. *The Feynman Lectures on Physics*. Boston: Addison-Wesley.

Sprott JC. 2003. *Chaos and Time-Series Analysis*. Oxford, UK: Oxford University Press.

Weinberg S. 1993. *The First Three Minutes: A Modern View of the Origin of the Universe*. New York: Basic Books.

FURTHER READING

Gleick J. 1987. *Chaos, Making a New Science*. New York: Viking.

A historical account of the development of chaos theory is described in this book.

Weisstein EW. Logistic map. MathWorld—A Wolfram Web Resource. http://mathworld.wolfram.com/LogisticMap.html

A detailed discussion of the logistic difference equation is given in this online source.

INDEX

Thinking About Equations: A Practical Guide for Developing Mathematical Intuition in the Physical Sciences and Engineering, by Matt A. Bernstein and William A. Friedman
Copyright © 2009 John Wiley & Sons, Inc.

Printed and bound by CPI Group (UK) Ltd, Croydon, CR0 4YY

27/10/2024

14580257-0003